Australian
BIRDS

Published by
Woollahra Sales and Imports
Unit 6, 32–60 Alice Street, Newtown, New South Wales, Australia, 2042
Phone: (02) 9557 8299 Facsimile: (02) 9557 8202
Email: wsi@pacific.net.au

Produced for the publisher by
Murray David Publishing
Publishing Director: Murray Child
Marketing Director: David Jenkins
35 Borgnis Street, Davidson, New South Wales, Australia, 2085
© Photographs: Ken Stepnell, 2004
© Text: Murray David Publishing Pty Ltd, 2004
Text by Dalys Newman
Printed in Indonesia

ISBN 1 876553 34 0

Australian
BIRDS

PHOTOGRAPHY BY KEN STEPNELL
TEXT BY DALYS NEWMAN

WOOLLAHRA

INTRODUCTION

Australia's vastness and its diversity of habitats has led to more than 700 species of birds making this continent their home. About 570 are resident breeding species, while the others are either regular or occasional non-breeding visitors.

Fossil records show that birds have been present in Australia for at least 130 million years. A number of families such as emus, mound-building birds, frogmouths and parrots seem to have originated in the southern super-continent of Gondwanaland, evolving from ancestors present when the Australian continent started drifting northwards, about 100 million years ago. As the continent drew closer to Asia, during the last 20 million years, other species from the north were able to colonise Australia by island-hopping.

Whatever their history may have been, many of Australia's birds have been on the continent for so long that they are quite unlike birds anywhere else in the world.

Birds have inhabited all corners of this vast continent: the rocky coastlines, wetlands, forests of all types, treeless plains, vast sandy deserts, mountains and hill country, fertile farmlands and teeming cities.

This rich variety of birdlife is made even more fascinating by the way most of them have physically adapted to special circumstances, for example the honey-eaters with their brush-like tongues for sucking honey from native blossoms, the parrots whose powerful bills enable them to crush and tear hard objects, and the mound-building birds such as the mallee fowl and brush turkey, whose strong legs and feet enable them to build the mound in which their eggs are incubated.

Encroaching settlement and civilisation has led to mass destruction and alteration of bird habitat. Feral mammals such as foxes and cats pose an additional threat to birdlife, and introduced birds compete with native species for food and nest sites. There are fears and doubts about the survival of many species, but hopefully, a stronger awareness of the dangers of habitat destruction will retain Australia's glorious wealth of birdlife.

Title page: The elegant white-browed woodswallow (*Artamus superciliosus*) sharing nest duties. **Above:** A mallee fowl (*Leipoa ocellata*) works his incubating mound. **Opposite:** Pelicans (*Pelecanus conspicillatus*) adrift.

Opposite, right and below: Australia's best-loved bird, the laughing kookaburra *(Dacelo gigas)* raises a wild chorus of crazy laughter at dawn and dusk throughout bushland and suburban areas. When in the throes of laughter they have a distinctive stance, tilting their heads skywards and flicking their tails up and down. Able to discern movement from afar, they make spectacular long swoops to seize their prey which includes lizards, snakes, mice, frogs, insects and the young of other birds. They lay 2-4 eggs in debris at the bottom of holes in tree hollows. One of the world's biggest kingfishers, the laughing jackass, as it is also known, reaches 470 mm and has a large head and oversized, extremely robust bill.

Opposite, above and overleaf: Second in size only to the ostrich, the emu *(Dromaius novaehollandiae)* is found on the plains and open woodlands throughout the Australian mainland. These huge, flightless birds usually travel in small flocks, walking long distances to find food and being capable of speeds up to 50 km per hour. Standing up to 2 metres tall, they have tiny wings and long, powerful legs. The female lays up to 20 large blue-green eggs in a scant nest on the ground, leaving the male to sit on them for about eight weeks and then care for the distinctively striped young for about a year and a half. The female may remain with the male and chicks or may leave and mate with another male.

Opposite: Large and colourful, the flightless cassowary *(Casuarius casuarius)* is found only in the dense rainforests of north-eastern Queensland. Their heads are topped with a large horny helmet, used to fend off obstacles as they move through forest undergrowth. Their call is a combination of guttural coughing and shrill whistling and their diet consists of berries, fruits, vegetation and sometimes carrion. Notoriously bad-tempered, they are shy birds usually only seen at the jungle edge or in clearings during dawn and dusk.

Below: The malleefowl *(Leipoa ocellata)* is one of Australia's unique mound builders. The female lays 16-33 eggs in incubating mounds built by the male of decayed vegetation mixed with earth or sand. The chicks hatch underground after about seven weeks and dig their way to the surface.

Right: A ground dwelling bird, the stately Australian bustard *(Ardeotis australis)* inhabits open grasslands, woodlands and pastoral country in northern and central Australia. Their range has been much reduced by shooting, foxes and habitat destruction. Gregarious and partly nocturnal, they feed on seeds, fruits, rodents, reptiles and nesting birds.

Opposite: Often seen in Darwin Botanic Gardens, the rose-crowned pigeon *(Ptilinopus regina)* is a glorious tiny (200-230 mm) bird with rich yellow-pink underparts and a broad yellow tail-tip. Tropical fruit-eating pigeons, they build round platform nests in low growth or vines and lay one large, white egg.

Above: A large, plump pigeon, the common bronzewing *(Phaps chalcoptera)* is found in forest and bushland throughout Australia. They are ground-feeders and travel considerable distances at dawn and dusk to drink at creeks and other watering places. Its call is a low, mellow repetitive 'oom-oom-oom'.

Above: Found in coastal and inland bush country in south-eastern and south-western Australia, the brush bronzewing pigeon *(Phaps elegans)* is commonly seen feeding quietly under trees or shrubs. They build frail, saucer-shaped twiggy nests near or on the ground in which they lay two white eggs. It is smaller than the common bronzewing and has a higher pitched call.

Right: The wonga pigeon *(Leucosarcia melanoleuca)* has a monotonous, piercing 'wonga' call which can be heard up to a kilometre away. Found in rainforest and tall timber country in the eastern coastal and highland areas, this bird was considered a tasty dish by early settlers and its numbers have consequently been greatly reduced.

Opposite: Easily tamed, the plump little spinifex pigeons *(Petrophassa plumifera)* were hand fed near Alice Springs during drought years. Inhabitants of rocky, hilly, spinifex country, near water, in inland and northern Australia, they run and dodge swiftly through terrain, flushing with sudden whirrs and flying with rapid wingbeats.

Above: The striking Torres Strait pigeon *(Ducula spilorrhoa)* migrates to Australia during the summer months to breed. Dense breeding colonies are found in mangroves or rainforest canopies in coastal north Australia and offshore islands. The mostly white bird often has brown fruit stains on its face and grows between 380-440 mm.

Opposite: Also known as the topknot, the crested pigeon *(Ocyphaps lophotes)* is found in open country where water is present as well as settled areas throughout the mainland. They are often seen on roadside posts or telephone wires and when in flight their wings make a distinctive whistling sound.

Opposite: One of the wonders of nature, the male satin bowerbird *(Ptilonoryhnchus violaceus)* builds a bower consisting of a platform of sticks with two parallel walls. He then proceeds to decorate its entrance with any blue item he can scavenge. This bower-building operation is allied to courtship and mating.

Left and above: The male and female regent bowerbirds *(Sericulus chrysocephalus)* are a disparate pair. The striking black and gold male far outshines his olive-drab mate. Found in the wetter parts of the Great Dividing Range and coastal areas of eastern Australia, these birds build small avenue-type bowers in secluded places.
The male, who is smaller than the female, keeps the area around his bower spotlessly clean, but inside is often untidy. He collects only a few display objects including shells, seeds, fresh leaves and coloured berries.

Right: The ringing call of the pied currawong *(Strepera graculina)* haunts lonely woodlands and snow-mantled high-country in eastern Australia. These large black birds with bright yellow eyes and white tail bases are particularly vocal in stormy weather. Bold and tame around settlements, currawongs are also commonly found scavenging in towns, orchards and pastures.

Below: Australian magpie *(Gymnorhina tibicen)* chicks grow up in a bowl-shaped nest of sticks and twigs, lined with grass, rootlets, wool or hair. The nests are usually in a tree fork 5-16 metres or more high. Found throughout Australia, in both cities and bush, these birds have a beautiful rich, mellow, flute-like call, heard at its best in late winter and early spring.

Opposite: A gregarious bird, the little raven *(Corvus mellori)* often lives side by side with the slightly larger Australian raven. A bird of the inland, it is found throughout Australia in open timber, dry scrub, mulga country and arid plains. Carrion eaters, ravens, despite contrary belief, are beneficial to farmers as they assist in wiping out large numbers of blowflies and bacteria.

Opposite, left and above: Australia's largest bird of prey, and the fourth largest eagle in the world, the wedge-tailed eagle *(Aquila audax)* can be seen soaring above diverse habitats from mountain forests to treeless plains and even cities throughout the country. Huge, powerful birds, their wingspan has been recorded at 2.85 metres. When not gliding high in the sky these birds are often seen at rest on trees, inland telephone poles or on the ground. They build huge, sturdy nests lined with fresh eucalypt leaves high in the trees and hatch up to three chicks. Old nests are used as feeding platforms. The juvenile birds are slow to acquire adult plumage and, as they reach adult size quickly, it is possible to see two birds of the same species showing entirely different colours.

Above and opposite: Young swamp harriers *(Circus aeruginosus)* are found in nests on the ground among rushes, dense grass or crop vegetation. The female incubates the eggs and feeds the chicks while the male hunts for food. These large, slim hawks are best known for their low sailing flight on upswept wings over open country throughout coastal Australia. Feeding on small mammals, ground-dwelling birds, reptiles and amphibians, these birds scream loudly when they attack their prey, probably as a 'scaring' tactic.

Right: The noisiest of all Australian birds of prey, the brown falcon *(Falco berigora)* has a repertoire of screeches and demented cacklings, at times sounding like a laying hen. Bold and rowdy hunters, their prey includes small animals and reptiles, birds and insects. These birds avoid nest-building when they can, using the deserted nests of crows or other hawks.

Opposite: The tawny frogmouth *(Podargus strigoides)* is a master of camouflage, assuming a stick-like mobility and blending into the branches of a tree. These birds are mainly ground feeders, eating insects, small vertebrates and lizards. The bird's bill is broad and flat, capable of opening widely, like a frog.

Above: The unmistakable, dog-like 'wook-wook' of the barking owl *(Ninox connivens)* is one of the most notorious night sounds of the Australian bush. During the winter breeding season, this robust owl also emits a piercing, sobbing scream. An efficient hunter, it feeds on rabbits and small hares as well as rats, mice, birds and other small animals.

Above and opposite: So-called because of their habit of roosting in barns, barn owls *(Tyto alba)* are nocturnal creatures floating like ghosts through savannah country and farmlands throughout Australia in search of rodents which they are able to locate by sound alone. They nest on decayed debris in hollow tree-trunks hatching up to six chicks. If flushed from their hollow trees or heavy foliage they are defensive but still extremely helpless in the face of daytime attacks and harassment from other birds.

Opposite: Bold and aggressive, the white-eared honeyeater *(Lichenostomus leucotis)* has been known to alight on people's heads to remove hair for nest-building. Its nest is a deep cup of bark shreds and grass, bound with spiders' web and lined with wool, fur or hair. A bird of varied habitats, it dines on a large number of insects as well as honey.

Below: Victoria's faunal emblem and one of the world's rarest birds, the helmeted honeyeater *(Lichenostomus cassidix)* is found only in a small creek-side colony of less than 200 east of Melbourne. Belligerent and vocal, these birds forage in foliage, hang upside down and often descend to the undergrowth or water to bathe. Their nests are deep, untidy cups of grass, bark, ferns and leaves.

Left: The inquisitive and aggressive blue-faced honeyeater *(Entomyzon cyanotis)* mobs owls and goannas, raids fruit trees, takes food in poultry yards and scavenges at rubbish dumps. Its nest-building habits are slovenly and temporary and it often borrows the nests of babblers, mudlarks and apostle birds. Found in open forest, it has a strident, penetrating call.

Opposite: Found in the rainforests of north-eastern Queensland, the yellow-spotted honeyeater *(Meliphaga notata)* eats fruits and berries as well as the usual nectar and insect diet. Its nest is a delicate cup of bark shreds, bound with spiders' web and decorated outside with pale-green lichen.

Above: A familiar sight in suburban gardens in eastern and south-eastern Australia, the yellow-faced honeyeater *(Lichenostomus chrysops)* favours blossoming shrubs. They live in the forests in spring and summer, forming into flocks as the weather cools and migrating to the north. It is on these migratory journeys that they pay their visits to home gardens.

Opposite: The singing honeyeater *(Lichenostomus virescens)* hardly does justice to its common name as its call is a scratchy, peevish 'scree-scree'. Widespread in mainland Australia, they are belligerent, bossy birds, often chasing other birds from prime feeding grounds. Their nests are frail untidy cups of faded grass or plant stems slung from fine branchlets in dense scrub.

Above: Like many birds, the New Holland honeyeater *(Phylidonyris novaehollandiae)* has adapted well to altered environments, finding food where it can in the suburbs. Found throughout south-eastern and south-western Australia it frequents banksia groves, feeding on the blossom and using the velvety down of the cone for nest lining.

Overleaf: The bush stone-curlew *(Burhinus magnirostris)* lays its two stone-coloured, brown and grey blotched eggs on the bare ground in open forest. Found in woodland and sandy scrub near beaches throughout Australia, they are active at night when their eerie, mournful call can be heard. Aborigines named this bird 'willaroo' after the sound of its call.

Opposite: The jewel-like Gouldian finch *(Erythrura gouldiae)* is better known in captivity than in the wild. Much exploited by bird-catchers, it is found in vegetation near water in the far northern areas of Australia. The bird was named by John Gould who wrote: 'I ... dedicate this lovely bird to the memory of my late wife who for many years laboriously assisted me with her pencil.'

Above: Found in tropical heathland and vegetation near watercourses, the star finch *(Neochmia ruficauda)* is also known as the red-faced or red-tailed finch. These birds are often seen in huge flocks, flying swiftly and making simultaneous twists and turns, like starlings. They lay up to seven white eggs in a domed, rounded nest.

Opposite: Sometimes seen in home gardens, the diamond firetail *(Emblema guttata)* builds bulky, bottle-shaped nests with a long entrance spout. Constructed of grass, they are lined with finer grasses, feathers and down. Up to nine white eggs are laid and both birds share nest duties.

Above: The red-browed finch *(Emblema temporalis)* is a tame, friendly bird, often seen at bird-tables in parks and backyards. Commonly seen in eastern and south-eastern Australia, they are also found on Kangaroo Island. Their large bottle-shaped nests are built in dense low scrub, prickly bush, orchard citrus, cypress and pines.

Opposite: Distantly related to starlings, the white-browed woodswallow *(Artamus superciliosus)* is a migratory bird, flying from eastern and inland Australia to nesting localities in south-eastern Australia where it builds small, bowl-shaped nests of grass and fibre strips.

Above: A skilled aerialist, the dusky woodswallow *(Artamus cyanopterus)* is capable of both long and high flight. In common with all woodswallows, these birds have a curious habit of roosting in clusters like a swarm of bees. Up to 50 gather on the trunk of a tree, later arrivals gripping the wings or shoulders of the birds beneath them.

Right: Tree martins *(Cecropis nigricans)* are found nesting in tree hollows throughout Australia. The birds choose a ready-made hollow in which to build nests of grass and leaves and partly close off the entrance with mud packing. These small, graceful birds are often seen in large flocks on summer evenings in both open timber country and settled areas.

Opposite and left: The male superb blue wren *(Malurus cyaneus)* is an eye-catching member of the wren family unlike his mate, the Jenny wren (above), who has far less glamorous plumage. In autumn and winter matters are equalised when the male moults into a less colourful eclipse plumage. These delightful little birds with their long cocked tails are found throughout Australia, in suburban parks and gardens as well as in natural bush haunts. When breeding, males and females persistently battle their own reflections in any shiny surface. Their brisk, merry trilling song can be heard from the tops of fences, thickets and shrubs.

Opposite: The distinctively shaped nest of the rufous fantail *(Rhipidura rufifrons)* is built out of fine strips of bark, moss and fine grass, bound externally by spiders' web. Found in coastal northern and eastern Australia, they frequent the dense, damp undergrowth of tropical rainforests and scrubs, monsoon forests and mangroves.

Above: Like most fantails, the willie wagtail *(Rhipidura leucophrys)* is rarely seen at any great heights, preferring to flit around lower tree branches or near the ground. Friendly and adaptable, this cheeky bird builds his distinctive nest in all manner of places, from fallen branches to street lamps and clothes hoists. He is often seen in the company of magpies and mudlarks.

Opposite: A member of the honeyeater family, the noisy miner *(Manorina melanocephala)* is often seen in suburban gardens where flocks will chase and harass dogs, cats and other birds. Found throughout eastern Australia, it nests in loose colonies building untidy, large, cup-like nests of twigs and grass, lined with fine grass, rootlets, hair, fur or wool.

Above: A superb mimic, the olive-backed oriole *(Oriolus sagittatus)* inhabits the forests and mountain slopes of northern and eastern Australia. Entirely arboreal, they feed on fruit, berries and insects and construct deep nests of bark strips, leaves, grass and moss, mostly among hanging outer foliage.

Left: So-called for its stiff, forward-curving bristles growing from near the base of the bill, the rufous bristlebird *(Dasyornis broadbenti)* is the largest member of this species. Shy and elusive, it has a silvery clear call and is more often heard than seen. It builds a large, rough nest in small tea-trees or clumps of sword-grass.

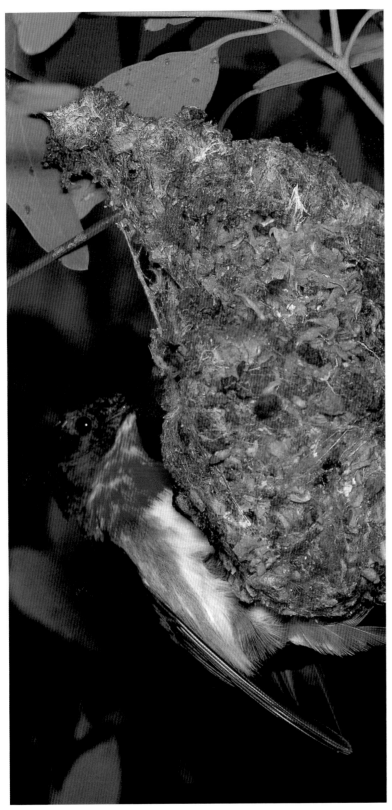

Opposite: The ground-frequenting white-fronted chat *(Ephthianura albifrons)* inhabits low ground-cover in the coastal heaths, plains, saltbush and samphire of south Australia. Chats are well known for their 'broken-wing' display, where they pretend to have a damaged wing to lure intruders away from their nests.

Left: A member of the Flowerpecker family, the mistletoebird *(Dicaeum hirundinaceum)* is a small colourful bird with a stubby tail, fine bill and tongue adapted to nectar-eating. However, its main diet is mistletoe berries and to deal with this diet it has a specialised alimentary system with a practically non-existent gizzard. They build fragile, flimsy nests out of wool, cobweb, down and flower petals, decorated with brownish materials.

Below: The grey butcherbird *(Cracticus torquatus)* derives its name from its habit of wedging larger prey in tree forks or fence wires and then dismembering it before eating. Found in forests, scrubland and suburbia in most parts of Australia, these birds have a beautiful piping song and build untidy saucer-shaped nests in saplings or scrubby trees.

Above: Found in open timber and savannah country throughout mainland Australia, the female hooded robin *(Melanodryas cucullata)* nearly always builds her nest on or near dead wood. The male is recognised by his distinctive pied plumage. These birds are exponents of the 'broken' wing act, hobbling along to lure intruders away from their nest.

Opposite: Both male and female scarlet robins *(Petroica multicolor)* share nest duties, the male usually feeding the female who then feeds her young. Also known as robin redbreasts, these distinctive birds are found in south-eastern and south-western Australia, building nests decorated with strips of bark and lichen in varied places such as behind loose bark and in sheltered cavities of charred trees.

Above: Commonly known as diamondbirds, the tiny, jewel-like spotted pardalotes *(Pardalotus punctatus)* are hard to see, moving like mice high in the treetops. These birds prefer to nest in tunnels, which they drill into the bank of a watercourse or roadside cutting. Nests have also been found in such places as sandpits, downpipes, corrugated iron roofs and hanging baskets.

Opposite: The dowdy but charming jacky winter or brown flycatcher *(Microeca leucophaea)* is found throughout most of mainland Australia. One of the most loved of the country's birds, it travels alone or in pairs and builds nests so small that the sitting bird all but covers it. Their dawn song is a clear and beautiful musical variation of the phrase 'peter-peter-peter'.

Opposite: The gorgeously coloured varied lorikeet *(Psitteuteles versicolor)* frequents flowering trees in northern Australia and coastal islands. It has a less strident call than other lorikeets but is an aggressive bird, chasing other birds who attempt to feed in the same trees. In flight its green wing-linings contrast with its blackish flight feathers.

Above: The only lorikeet with an unmarked green head, the greenie or scaly-breasted lorikeet *(Trichoglossus chlorolepidotus)* is found in euca-lypt woodlands and forests throughout coastal eastern Australia. In common with all lorikeets, these birds perform gymnastic feats when feeding. They are often found in association with rainbow lorikeets.

Left: Small and dainty, the purple-crowned lorikeet *(Glossopsitta porphyrocephala)* is the only lorikeet found in the south-west of western Australia. It is a nomadic bird, its route following the seasonal blossoming of the gum trees. A quiet feeder, it crawls like a mouse through leaves and blossom, occasionally uttering quick 'zit-zit' notes.

Previous pages: Rainbow lorikeets *(Trichoglossus haematodus)* are easily tamed when provided with nectar in gardens. These playful birds inhabit virtually all types of timbered country in coastal eastern and south-eastern Australia and are often seen in city parks and gardens. They feed on fruit and blossoms, chattering and screeching incessantly.

Opposite: Found only in Tasmania and the Bass Strait Islands, the green rosella *(Platycercus caledonicus)* inhabits dense mountain forests to lightly timbered farming country. In winter, large flocks gather in hedgerows — they thrive on the introduced hawthorn bush.

Above: Frolicsome and friendly, crimson rosellas *(Platycercus elegans)* thrive in National Parks, camps and reserves. Young rosellas congregate in small flocks, whilst mature birds are usually seen in pairs. They inhabit rainforests, fern gullies and coastal scrub and feed on seeds, nectar, soft fruits and some insects.

Above: Eastern rosellas *(Platycercus eximius)* make their nests on decayed debris in tree hollows, stumps or hollow fence posts. These extremely colourful birds are a common sight in south-eastern Australia and Tasmania, preferring open forests and woodland. They have a strong, undulating flight, spreading their tail before swooping to land.

Opposite: The gaudily coloured male red-capped parrot *(Purpureicephalus spurius)* inhabits the jarrah and other tall trees of south-western Western Australia. Its projecting bill is especially suited to cracking the hard nuts of these trees. Quite common, it is sometimes seen in the suburbs of Perth. They feed silently but when alarmed utter a series of harsh, clanging shrieks.

Above: The migratory swift parrot *(Lathamus discolor)* breeds only in Tasmania and some Bass Strait islands, arriving in August and September and returning to the mainland between January and May. A fast and erratic flier, it has a tinkling, chattering, musical call which it sounds on the wing.

Opposite: Princess parrots *(Polytelis alexandrae)* are rare, little-known inhabitants of the arid interior. Highly nomadic, their movements are governed by the flowering of acacias and availability of watercourses. They have an easy, undulating flight and are often seen perching lengthways on tree limbs. These parrots are quiet and seldom call.

Above and opposite: A bird of the rainforests and jungles of the eastern Cape York Peninsula, the eclectus parrot *(Eclectus roratus)* belongs to a genus found mainly in islands north of the continent. They are strong fliers and on long flights to and from roosting trees they fly high above the forest canopy. Noisy and conspicuous, their raucous screeching and squawking is a familiar rainforest sound. Unlike most other species of parrot, the sexes of this bird are very different, the male not as colourful as his mostly scarlet mate.

Above and opposite: The red-winged parrot *(Aprosmictus erythropterus)* is found in varied habitats in northern and inland eastern Australia. Often seen in pairs or small flocks, they are wary and difficult to approach. When disturbed they rise into the air and, calling loudly, fly off to the next grove of trees. When courting, the male, chattering softly, flies in circles around the female. The immature bird, whatever sex, has the same plumage as the female. The adult male is bright pale-green with a prominent scarlet shoulder; the female is mostly grass green with a thin scarlet band across her wing.

Above: Also known as the grass parrot, the red-rumped parrot *(Psephotus haematono-tus)* is found mostly in inland south-eastern Australia. Seldom far from water, they inhabit grasslands, open woodlands, farmlands and suburban gardens. They are friendly and will walk or waddle along the ground rather than take to their wings. Their call is a pretty warble, unusual for a parrot.

Opposite: Not as active or noisy as most parrots, the brilliantly coloured king parrot *(Alisterus scapularis)* inhabits rainforest, open woodland and dense scrubland along Australia's eastern coast. These birds prefer large trees and timber-milling has severe-ly restricted their range.

Opposite: Bourke's parrot *(Neophema bourkii)* is a bird of the mulga country and other inland acacia scrubs. A small softly coloured grass parrot, it has regular drinking habits, going to watering places before sunrise and at dusk.

Above: One of the world's most popular cage-birds, the budgerigar *(Melopsittacus undulatus)* is probably the most plentiful parrot in Australia. They form into huge flocks, flying in tight, precise formations, twisting and turning as one and often darkening the sky. Extremely nomadic, their movements are governed by the availability of water and seeding grasses.

Above: The long-billed corella *(Cacatua tenuirostris)* is slightly larger than the little corella, with a much longer whitish bill. It inhabits the open forests and woodlands in the higher rainfall areas of southern Australia and is seldom found far from water.

Opposite: The beautiful pink or Major Mitchell's cockatoo *(Cacatua leadbeateri)* is found mainly in arid or semi-arid woodland, including mallee. A relatively rare bird, it has declined in numbers in eastern states due to destruction of habitat.

Above and opposite: The noisy, conspicuous sulphur-crested cockatoo *(Cacatua galerita)* is found in pairs or small family parties during the breeding season and, at other times, in flocks of sometimes hundreds of birds. Found in a wide variety of habitats they have been troublesome pests in cereal-growing areas.

Overleaf: Large flocks of little corellas *(Cacatua sanguinea)* can often be seen covering trees by water or in white clouds on the ground. Ground-feeders, these nearly crestless cockatoos eat seeds from surface plants and dig for bulbs and roots. Widespread throughout most of Australia except the southern areas, they are troublesome pests in rice fields.

Opposite: The somewhat owl-like gang-gang cockatoos *(Callocephalon fimbriatum)* frequent mountain country in NSW and Victoria in summer, moving down to the plains in winter. Their call is unique — a prolonged creaky growl ending with an upward inflection. The male has a bright red head and untidy crest while the female is mostly grey.

Above: The only smallish parrot in Australia with a crest, the cockatiel *(Nymphicus hollandicus)* inhabits open inland country and is often seen by the roadside and on telephone wires. Its call is a melodious, loud, rolling chirrup, ending with an upward inflection.

Above: One of the most common and widespread of Australia's birds, the galah *(Cacatua roseicapilla)* mates for life, often returning to the same nest site year after year. They spend many hours feeding on the ground, moving over it with an awkward rolling gait, but when they take to the skies they are an impressive sight with the sun highlighting first their grey-pink underparts and then, as they wheel, the soft grey of the back and wings. After their evening drink they move in small groups toward roosting trees.

Opposite: Northern Australia is the stronghold of the red-tailed black cockatoo *(Calyptorhynchus magnificus)* where flocks of up to 200 can be seen in dry woodland or trees bordering watercourses. The male's striking scarlet tail-panels do not reach their full glory until the bird is four years old. On moonlit nights they often fly about uttering a call which sounds like a large rusty windmill.

Above and opposite: One of the world's rare waterfowl, the Cape Barren goose *(Cereopsis novaehollandiae)* breeds only on islands off southern Australia, several of which have been proclaimed reserves. There are thought to be only between 6000-8000 of these birds in existence. Grazing birds, they seldom enter the water, although they are naturally strong swimmers. Wary and hard to approach, they have a powerful flight, accompanied by a harsh clamour. Nests, in which 6 – 8 dull white eggs are laid, are built of twigs and grass on the ground beside a bush, tussock or rock.

Opposite and above: Best known for their spectacular dancing, groups of brolgas *(Grus rubicundus)* can often be seen, accompanied by loud trumpeting calls, leaping, bowing and high-stepping in riotous display. The most widespread of the country's two cranes, brolgas inhabit areas beside marshes and ponds, large flocks being found in northern Australia. They construct nests of grasses and plant stems on small islands in swamps and occasionally on bare ground. Two white, brown and lavender spotted eggs are laid.

Opposite, left, below and overleaf: Black swans *(Cygnus atratus)*, the country's only native swans, are highly nomadic birds found throughout Australia. They have been adopted by Western Australia as the state emblem. A familiar sight on ornamental lakes, these birds breed freely and often become tame. After breeding they moult and become temporarily flightless. Nests, in which 4-7 greenish white eggs are laid, are a large heap of reeds, grasses and weeds and are constructed on islands and in shallow water. These birds are found on either fresh or brackish water and their call is a far-carrying musical bugle, uttered on water or in flight.

Opposite, right and below: Perching on boats and wharf piles or gliding across calm waters, the Australian pelican *(Pelecanus conspicillatus)* is one of the most evocative of all waterbirds. Found on the margins of coastal or inland waters, they feed in a close circle, driving the fish into shallow water and then lunging with their large bills into the school of fish. Large birds, with a wingspan of 2.5 metres, they have enormous bills and bill pouches. These are used in securing food as well as for cooling the bird in hot weather when they open their bills and rapidly flutter the skin. Graceful in flight and on the water, they have a waddling gait on land due to their widely set apart legs. Nesting colonies are large and usually remote. Nests, in which 2-3 dull white eggs are laid, start as a scrape and are progressively lined with sticks, grass and seaweed.

Above and opposite: Found mainly on inland waters, the little pied cormorant *(Phalacrocorax melanoleucos)* is the smallest of Australian cormorants. These birds usually fish alone but roost and nest in groups. When in flocks, cormorants fly in changing formations, giving rise to the myth that they form letters in the sky. They have sparse plumage which easily becomes waterlogged and spend much of their time drying their wings on piles, piers, rocks and buoys. Excellent fishermen, their diet includes small fish, eels, frogs, crustaceans, aquatic insects and occasionally ducklings and small mammals.

Right: Also known as the snake-bird because of its long, sinuous neck, the darter *(Anhinga melanogaster)* inhabits larger shallow waters, both fresh and salt, throughout mainland Australia. They also have a snake-like style of swimming, with their body submerged and only the slender head and neck visible. Unlike cormorants, who are seldom very vocal, darters have loud, rowdy cackles.

Opposite: Often seen in large flocks made up of mated pairs, the plumed whistling duck *(Dendrocygna eytoni)* camps by water during the day and, when moving out to grasslands at night, flies with incessant whistling. Although mainly tropical they have been sighted near southern waters. They often dabble, but do not dive to feed, sometimes grazing on areas far from water.

Above: The black duck *(Anas superciliosa)* is the typical Australian 'wild duck' and is found throughout the country on waters of all kinds. It feeds by dabbling or up-ending on a diet including plankton, aquatic plants, grasses, seeds, insects, small crustaceans and molluscs. Although quite docile in inhabited areas, it is difficult to approach in the wild.

Right: A decidedly strange creature, the male musk duck *(Biziura lobata)* has a large leathery flap under his bill and stiff, pointed tail feathers which he fans forward over his back during courtship displays. These ducks dive for food, swimming mostly submerged like a cormorant and often doze on the water's surface with their tails fanned out.

Opposite: The banded plover *(Vanellus tricolor)* is also known as a lapwing because of its quick, clipped flight. It feeds mainly on insects and is found on poorly grassed paddocks and plains throughout Australia except for the far north. When disturbed, they run quickly in short bursts, then stand perfectly still. Their nests are small depressions on the ground.

Above: Found in wet grasslands and swamps, the masked plover *(Vanellus miles)* is easily identified by its bright yellow facial wattle. Also known as the spurwing plover, it has small yellow spurs on its shoulders which are used for fighting and defence. Noisy birds, they have a strident call and can be aggressive when nesting, diving at intruders.

Opposite: A familiar waterhen of willow-fringed town lakes, the dusky moorhen *(Gallinula tenebrosa)* builds a bulky nest of sticks, bark and grass in vegetation near water on which to lay her 7 – 10 spotted eggs. She often raises two clutches of young in one season.

Above: The swamphen *(Porphyrio porphyrio)* builds its nest and resting platforms out of trampled reeds or cumbungi, laying 3 – 5 sandy, chestnut spotted eggs. Swamphens have an unusual feeding method, using a foot as a hand and holding the food firmly between the toes, similar to a cockatoo. Their call is a series of loud, hacksaw-like screeches.

Above: Often seen feeding in paddocks alongside cattle, the cattle egret *(Ardeola ibis)* is the smallest of Australia's egrets. A sprightly bird when catching insects disturbed by grazing stock, at rest it seems chunky, hunched and deep-jowled. These birds are relatively new to the country, being first sighted here in the early 1900s.

Opposite: The large egret *(Egretta alba)* is found in river shallows, swamps and lagoons throughout the mainland and Tasmania. By far the tallest egret in the country, its head and neck when stretched out are one-and-a-half times as long as its body. They forage in the water for food, remaining motionless for long periods of time before making a lightning strike at their prey.

Opposite: The stately, colourful jabiru or black-necked stork *(Xenorhynchus asiaticus)* is Australia's only representative of the stork family. Widespread throughout the country, they are tall striking birds with a standing height of 1.2 metres, half of which is made up by their long legs. They fly with necks and legs extended and forage in swamps, tidal areas, woodlands and grasslands.

Right: Often seen in pairs or small flocks, the royal or black-billed spoonbill *(Platalea regia)* is a spotless white bird with black bill and legs. They wade in shallow water, sweeping and fossicking with the partially open sensitive tips of their bills in search of aquatic animals. The spooned bill measures 20 cm in length.

Above: Magpie or pied geese *(Anseranas semipalmata)* are sometimes seen in huge flocks ranging from 50 000 to 100 000 in number. Once widespread throughout the country, drainage of swamps, grazing, shooting and poisoning has limited its area to the coastal floodplains of tropical northern Australia. Attempts are being made to reintroduce them to other areas.

Opposite: A valuable insect destroyer, the straw-necked ibis *(Threskiornis spinicollis)* is also known as the Farmer's Friend for its ability to keep the grasshopper and locust population under control. This handsome bird has a straw-like tuft of breast plumes and is found in swamps, irrigated areas and grasslands throughout the country.

Right and below: White ibis *(Threskiornis molucca)* form large breeding colonies in areas with assured food and water supplies. Nests are compact shallow cups of reeds, sticks and bushes built over water in dense trees, trampled swamp growth or mangroves. Neighbouring nests become trampled into a continuous platform. They lay 2-4 white eggs, which quickly become dirty. Also known as the Sacred Ibis, this bird is very similar to the sacred bird of the ancient Egyptians whose image was painted on walls and papyrus.

Opposite, left and below: The little or fairy penguin *(Eudyptula minor)* is the world's smallest penguin and the only one to breed in Australia. At their breeding colony on Phillip Island, in Victoria, the birds are a popular tourist attraction. Every evening, during a particular stage in their breeding cycle they waddle ashore to the burrows they vacated in the morning for their fishing expeditions at sea. Their nests are built either in small hollows under bushes or in burrows 1.5 metres deep. Eggs are laid in early spring, the adults taking turns incubating them and going to sea to fish. Eight weeks after hatching, the parents desert their young and soon after the chicks leave for the sea. Powerful swimmers, penguins literally 'fly under water' propelled by their stiff, flattened flippers and using their webbed feet and small tails as rudders. They swim low in the water, often only with their head and tail visible, hunting for fish, squid and crustaceans.

Above and opposite: Spectacular divers, Australian gannets *(Morus serrator)* will plunge into the sea from great heights, 20 metres or more, hitting the water with an impressive splash. Found in the coastal waters of southern Australia, they nest in colonies, building mounds of seaweed and vegetation, held together by hardened excreta. The immature bird looks very different from the adult, having grey-brown, 'salt-and-pepper' plumage.

Above: Not usually seen on seashores, the whiskered tern *(Chlidonias hybrida)* is found on freshwater swamps, lakes, irrigated pastures and sewage farms throughout mainland Australia. They are graceful in flight, taking food by plunging below or skimming the water's surface.

Opposite: Also known as sea-swallows, fairy terns *(Sterna nereis)* are found along the western and southern coasts of mainland Australia and around Tasmania. Noisy birds, they nest in small colonies on offshore islands and sandy coastal inlets. They forage for food by flying low over the water, head down, periodically plunging in to feed on small fish.

Opposite: The big, solid Pacific gull *(Larus pacificus)* reaches 66 cm in height, with a very deep, heavy 6-cm bill. They will carry tightly closed shellfish to a height of 10 metres and then drop them to the rocks below to crack them open. In high winds these gulls sail majestically, often following fishing boats and ships. They are aggressive and highly vocal.

Above and overleaf: Probably the country's most prolific bird, the short-tailed shearwater *(Larus novaehollandiae)* is seen in immense flocks off south-eastern Australia during the summer months. The muttonbird of Tasmania, these birds have been commercially exploited. However, the industry is well controlled, only the young birds being taken and the eggs and adults strictly protected.

Left and above: The jewel-coloured sacred kingfisher *(Halcyon sancta)* is found throughout most of Australia except for the drier inland areas. They inhabit forests and woodlands, usually not far from water, and the margins of lakes and rivers, mangroves and seashores. They hunt on both land and in shallow water and build nests on debris in tree hollows, in holes in termite mounds and occasionally in tunnels in river banks. Very noisy in the breeding season, their call is a clear, measured 'dek dek dek'.

Above and right: The ornately coloured rainbow bee-eater *(Merops ornatus)* is so-called for its ability to remove the stings of bees or wasps before swallowing them. Great insect-catchers, they are skilled aerialists, catching all manner of flying insects on the wing. These birds drill up to 1.6 metre long tunnels in

sandy cliff faces or banks in which to build their nests. Invariably the hole entrance is directed away from prevailing winds, so that dust and rain cannot blow in. This species is migratory and nomadic, wintering in the north and returning to the south to breed in the spring.

Overleaf: Silver gulls *(Larus novaehollandiae)* are a familiar sight to most Australians. Seen both at the seashore and far inland, gracefully wheeling above newly ploughed fields and scavenging at rubbish dumps, they have adapted well to urban environments. They nest in colonies on offshore islands, laying pale brown, thickly blotched eggs. The chicks are slow to acquire adult plumage, the immature birds having dark bills and legs and mottled feathers.

INDEX

PEOPLE
WHO CHANGED THE
WORLD

PEOPLE
WHO CHANGED THE
WORLD

igloobooks

Published in 2014
by Igloo Books Ltd
Cottage Farm
Sywell
NN6 0BJ
www.igloobooks.com

A copy of the British Library Cataloguing-in-Publication
Data is available from the British Library

SHE001 0714
2 4 6 8 10 9 7 5 3 1
ISBN:978-1-78440-311-9

Written by Martin Howard

Printed and manufactured in China

CONTENTS

INTRODUCTION

Whether by accident or through careful planning, there have been people throughout time who stand out because, by changing events and circumstances, they have altered the course of history, sometimes even after their deaths.

Some of these people, such as Alexander Fleming and Marie Curie, made discoveries that changed medical procedures. The Wright brothers and Thomas Edison created inventions that changed our world forever. Others, such as John F. Kennedy or Martin Luther King, changed the course of history by brave actions and decisions. Art, Literature, Philosophy, Music and, more recently, the internet – every aspect of our world – has been changed in dramatic ways by people. The actions, inventions and impact of many of these people are explained in this book.

Muhammad Ali

Born 1942

American

Heavyweight boxer and sports ambassador

Considered by many to be the most outstanding sportsman of the modern age, Ali was born in Louisville, Kentucky, and christened Cassius Marcellus Clay. He first came to prominence in his chosen field of boxing during the 1960 Olympic Games in Rome, where as an amateur he won a gold medal in the light heavyweight division. Clay's fleetness of foot in the ring was amazing. As he famously said, he could "float like a butterfly and sting like a bee." A few months later he turned professional and, to the surprise of many, took the world heavyweight crown by beating Sonny Liston in seven rounds at Miami Beach, Florida, four years later. He was the first man to win the heavyweight title three times.

Clay became increasingly politicized during this period and joined the radical movement, the Nation of Islam, an African-American separatist group founded by Elijah Muhammad. Clay declared that his given name was that of a slave and that from then on he would be known as Mohammad Ali. Ali's support for the Nation of Islam caused outrage in some quarters and this intensified when he refused to join the army in 1967 at the height of the increasingly unpopular Vietnam War. Ali announced that his new religion meant that the war was really none of his business.

This decision had severe consequences for his boxing

Above: "The Greatest" went to London in 1966, where he trained at the Royal Artillery Gymnasium before his title defense against Henry Cooper.

career, as he was stripped of his heavyweight title, as well as being convicted on charges of evading the draft. Ali remained in the wilderness until 1970, when the US Supreme Court unanimously overturned his convictions. Despite this hitch, Ali returned to the ring in great form and he fought at his ruthless best during the 1970s. His pre-and post-match press conferences also showed him to be a brilliant speaker, one able to psychologically undermine most prospective opponents.

During the next 10 years or so Ali took part in some of the greatest boxing matches of all time, getting into the ring with some of the leading heavyweights of the age. There were three unforgettable bouts with Joe Frazier, to whom Ali lost his title for the first time in 1971. There was the classic title victory over George Foreman, the "Rumble in the Jungle", in Zaire during October 1974, and a pair of championship fights with Leon Spinks, to whom Ali lost his title in February 1978, but then regained it for the second and final time the following September.

Ali finally retired from the ring after being bested by the relatively unknown newcomer, Trevor Berbick, in 1981, but he did embark on a successful post-professional career. Three years later he was diagnosed with the incapacitating Parkinson's syndrome, a progressive degenerative condition that affects the brain function. Despite this diagnosis, Ali has striven to lead an active life. He undertakes charitable work for the Nation of Islam and continues to make popular guest appearances, not least on talk shows where his wit and erudition enthrall audiences. He also supports a great many fundraising events to help disadvantaged young sportsmen and women.

Above: The second Ali-Frazier fight took place at Madison Square Garden on January 28, 1974. Ali was the unanimous winner.

Above right: Ali received the Presidential Citizens Medal from President Clinton at the White House in Washington DC.

Right: Ali holding the Olympic torch during the Opening Ceremony of the 1996 Centennial Olympic Games in Atlanta, Georgia.

Kofi Annan

Born 1938
Ghanaian
International statesman

Kofi became the seventh secretary-general of the United Nations (UN) in 1997. The job of secretary-general is to negotiate peace, uphold security and find solutions to conflicts across the world. The UN works to promote human rights, and as part of that, fights terrorism and tries to alleviate the suffering caused by the HIV/AIDs epidemic, especially in Africa. Kofi Annan rose to the challenge of the job. Not only was he universally praised for his negotiating skills and vision, but in 2001 he was awarded the Nobel Peace Prize.

Kofi Annan was born in Ghana and after completing his education in Ghana and the USA, he joined the orld Health Organization, an agency of the UN. After a period as director of tourism for Ghana, Annan worked for the UN, gaining experience in human resources and financial management. In the early 1990s he was head of Peacekeeping Operations. He was highly praised for his work during the civil war in Bosnia, where he handled the tricky of transfer of peacekeeping operations that allowed the UN to withdraw its force.

Clearly dedicated to the founding principles of the UN, Annan was recommended as the successor to the Egyptian secretary-general Boutros-Boutros Ghali, who had held the post since 1992. An independent operator, Boutros-Ghali had alienated some member nations, notably the USA, so when Annan took over in 1997, he faced a tough job to win back the respect of the USA for the UN. His long years in administration enabled him to reform the budget of the United Nations and simplify some of it operations. He also set plans in motion to tackle the AIDs epidemic in Africa and reform human rights abuses throughout the world. This new show of efficiency and positive action impressed the USA, and Annan's obvious

Below: Kofi Annan shaking hands with members of press after his final press conference as Secretary-General in December 2006.

Above: Sahrawi refugees in Algeria wait for Kofi Annan while he makes another attempt to solve the Western Sahara conflict.

competence won him many admirers.

Annan trod a careful diplomatic path when dealing with the explosive situation in the Middle East. After the 1991 Gulf War, UN weapons inspectors had been assigned to trawl Saddam Hussein's Iraq looking for "weapons of mass destruction." Saddam did everything he could to obstruct the weapons' inspectors, but Annan was able to defuse a succession of crises at the turn of the century.

A few months after Annan was appointed for a second term, he had to deal with events that followed September 11, 2001, the day of the Al Qaida terrorist attacks on New York's Twin Towers, an act of violence that led to the "War on Terror." The resulting American invasion of Iraq in 2003 took place without approval from the UN, and Annan had to work hard to repair diplomatic relations with the USA, while at the same time leading the international efforts to fight terrorism.

Annan's term as secretary-general came to an end in 2006 after many successes. The Global AIDS and Health fund, which he established, has received some $1.5billion in contributions, and this has had a real effect on saving lives. His 2001 Nobel Prize praised him for "bringing new life to the organization" of the UN, and his efforts to make the organization run more efficiently were almost as important as his commitment to peace and security across the globe.

Below: Throughout his career, Kofi Annan demonstrated firm resolve and calm in the face of crisis.

Archimedes

c. 287–212 BC

Greek

Mathematician, physicist, inventor, and engineer

Archimedes remains the most famous mathematician of the Classical world and the scientist whose formulas relating to spheres and other solid objects remain at the basis of all mathematical knowledge.

Born in the city of Syracuse in what is now Sicily, little is actually known of Archimedes' early years. It is thought that he traveled to Egypt to study in the Greek-controlled city of Alexandria, home to what was the greatest library of the age.

Archimedes is popularly known for two things. First, for creating Archimedes' Screw, a brilliantly simple, but practical irrigation device. It consists of either a tube wound spirally around a cylindrical axis, or a cylinder enclosing a screw that forms a spiral chamber from top to bottom. The bottom of the cylinder can be dropped into water and as the screw is turned,

Above: The brilliantly simple "Archimedes Screw" is still used today for irrigation in many deprived regions of the world.

Left: The basis of all mathematical knowledge relating to the formulae of spheres and other solid objects is entirely owed to Archimedes.

the liquid is raised. The screw is still used today in developing countries around the world.

Second, Archimedes is remembered for his cry of "Eureka, Eureka!" ("I have it, I have it!") as he ran home naked from a public bath house. Asked to work out whether a solid gold crown contained any silver impurities, Archimedes found the solution to the problem in his bath, by noting how much water was displaced by his body when he sat down in the bath. "Archimedes Law" relates to the buoyancy of objects in water, and states that the weight of water displaced by an object is equal to the object's weight. A boat sinks into a lake until the weight of the water it has displaced is the same as the boat's weight.

However, Archimedes' true genius was very much broader, and his greatness really rests with his discovery of various mathematical rules that relate to the areas and volume of

Above: Archimedes invented a devise to focus the sun's rays—here he demonstrates how to direct the mirrors at the invading Roman warships.

Below: Archimedes was the pre-eminent mathematician of the Classical World. Unfortunately, all his astronomical work has been lost.

spheres and other figures. Extremely interested in mechanics, he was also responsible for establishing the principles of the lever. Nine of his treatises survive in Greek, and it is known from the writings of later mathematicians that several more works have disappeared in the 2,000 years since his death.

It is unfortunate that his work in the field of astronomy has been lost, but it is clear that he set new standards in mathematical rigor by combining a flexible approach to a problem and examining it from several angles.

Archimedes' death was apparently caused by his love of mathematics. In 213 the Romans besieged the city of Syracuse and Archimedes invented several machines to protect the city walls. He created huge grappling devices, or "tongues," that were fitted to the city's seaward walls and were apparently used to lift Roman warships out of the water and overturn them. Nevertheless, the Romans broke into the outer city in 212 BC during a local festival. So the story goes, a Roman soldier challenged Archimedes while he was concentrating on a mathematical problem. Receiving no reply, and thinking that the mathematician was snubbing him, the Roman killed Archimedes with his sword.

Neil Armstrong

1930–2012

American

Astronaut, pilot and aerospace engineer

With the celebrated words, "That's one small step for man, one giant leap for mankind," Neil Armstrong assured his place in history by becoming the first man to set foot on the Moon on July 20, 1969. Historians have since claimed that his supposedly spontaneous words were scripted, but nothing can lessen Armstrong's achievement.

Armstrong was born in Wapakoneta, a small town in central Ohio, and graduated from Purdue University, Indiana. He then joined the US Air Force, where he earned his pilot's wings, and saw active service in the Korean War (1951–1953) flying fast jet fighters. He was chosen to become a member of the National Aeronautical and Space Administration's (NASA) small, elite band of astronauts in 1962. Four years later, after intense training, he was selected to command Gemini 8. The Gemini program was the first stage in preparing for space flights to the Moon, and astronauts practiced such techniques as space walking, docking, and landings. The Gemini 8 mission was launched on March 16 1966 and, Armstrong with his co-astronaut David Scott successfully completed the first space docking and then made an emergency landing. The mission lasted a total of 10 hours, 41 minutes and 26 seconds.

The program that was to take men to the Moon was give the codename Apollo and the first mission took place in late January 1966. After several training missions NASA was read to attempt the Moon landing in mid-1969. Three astronauts, including Armstrong, were selected to crew Apollo 11. All were highly experienced. Aside from Armstrong's Gemini mission, the other two crewmen, Edwin "Buzz" Aldrin and Michael Collins, had both flown before – Aldrin on Gemini 12 on November 11, 1966 and Collins on Gemini 10 on July 18 the same year.

Apollo 11 blasted off from the Kennedy Space Center on July17, 1969, embarking on a mission that would last in total 8 days, 3 hours, 18 minutes and 35 seconds. The launch and journey to put the command module and attack lunar module in orbit around the Moon was completed by July 19 and the first landing on the satellite's surface took place on the 20th. Collins remained with the orbiting command module, leaving Aldrin and Armstrong to touch down on the Moon's surface in the lunar model. Armstrong became the first man to step down on to the Moon at the Sea of Tranquility at 10.56 pm EDT, and Aldrin followed shortly afterwards. The two astronauts spent around two hours exploring the immediate area, unfurling the Stars and Stripes flag and collecting various rock and other samples. The events were broadcast into televisions in millions of homes around the world.

The lunar module's lift-off, docking with the command module, the return flight to Earth, and the splash-down in the Pacific Ocean on July 24 went smoothly, and the three astronauts then spent some time in quarantine before receiving an ecstatic home-coming. Apollo 11 was Armstrong's last space mission and, after retiring from the program, he taught aerospace engineering at Cincinnati University (1971–1979). His autobiography, *First on the Moon*, was released in 1970.

Above: Neil Armstrong, Commander of the Apollo 11 Lunar Landing Mission posing for his official portrait in July 1969.

Above: The astronauts' motorcade passes through 42nd Street, Manhattan as they are welcomed home with a ticker-tape parade, August 15, 1969.

Right: The Apollo 11 Spacecraft launching from Kennedy Space Center July 16, 1969, with astronauts Armstrong (commander), Collins (command module pilot), and Aldrin.

Below: Neil Armstrong's footprint on the Sea Of Tranquility on the Moon's surface, where it has remained since July 20, 1969

John Logie Baird

1888–1946

Scottish

Electrical engineer and television pioneer

The Scottish electrical engineer, John Logie Baird invented the world's first working television. His broadcast system was the first that was capable of transmitting sound and pictures together. Television dominates almost every area of life today, but the initial idea of television – a small box showing endless free entertainment available within the home – was revolutionary.

Baird was born in the town of Helensburgh, an elegant Georgian town on the Firth of Clyde to the northwest of Glasgow. He completed his education at the Royal Technical College in that city, where he studied electrical engineering, and after his graduation took up a post with the local Clyde Valley Electrical Company. Baird did not stay with the company for too long as he was plagued by ill health and he eventually gave up his position. He next had a brief career as a sales representative but then moved south into England. He settled in Hastings, a quiet seaside resort on the southeast coast, in 1922.

While living in Hastings, Baird began significant investigations into television and it was in 1923 that he first demonstrated a televised image. He applied for a patent on July 26, stating that he had devised a "system of transmitting views, portraits, and scenes by telegraphy or wireless telegraph," and it was granted the following year. Baird's work continued and in 1925 he produced his "Televisor", a working television constructed out of, among other things, biscuit tins, darning needles and small tea chests. The next year he made the first public demonstration of television to the Royal Institute in London. He also patented the "Phonovisor", the earliest video recorder, which recorded images on a wax record-type discs. Baird also took time to patent a type of radio detection, a forerunner of radar that consisted of "a method of viewing an object by projecting upon it electromagnetic waves of short wavelength."

Below: John Logie Baird adjusting the transmitter of his new "wireless vision" system, which he called the "Televisor.

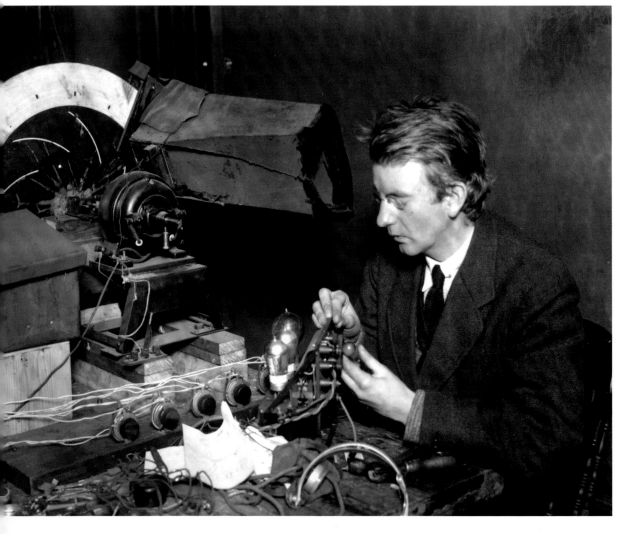

Above: In 1925 Baird's television was constructed from a collection of wires and disks, including biscuit tins, darning needles, and small tea chests.

Below: John Logie Baird showing off one of the most popular designs of prewar television sets which cost £35 (about $70).

Although the British Broadcasting Corporation (BBC) adopted Baird's television system in 1929, and the new improved version that gave a much clearer picture, in 1936, Baird had a rival. The Russia-born electrical engineer Vladimir Zworkin, who had emigrated to the United States in 1919, filed a patent in 1923 for what he called an "Iconoscope", which was a cathode ray television transmitter. The following year, he filed for a "Kinescope", a cathode ray television receiver. Both were essential elements of a television system. Baird initially held the advantage, as his device was basically a simple mechanical system, while Zworkin's electronic devices were held back by the slow pace of development in electronics. But Baird's equipment was slow and heavy, whereas Zworkin's seemed easier to handle, and the BBC eventually adopted the Zworkin system. It was further developed by the company Marconi-EMI in 1936.

Baird continued to make important contributions in the fields of electrical and mechanical engineering. He demonstrated the first electronic color television in 1944, two years before his death in 1946.

The Beatles

George Harrison 1943–2001
John Lennon 1940–80
Paul McCartney Born 1942
Ringo Star (Richard Starkey) Born 1940

British

Musicians

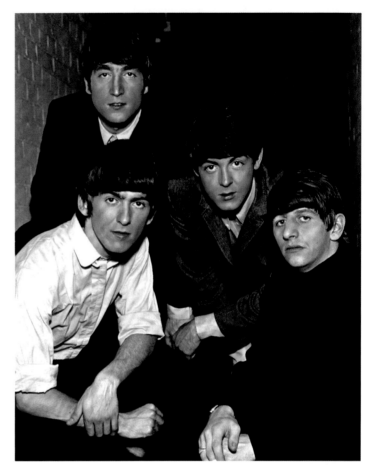

Above: In 1964 the Beatles became a global sensation unmatched by any other group of musicians before or since.

The Beatles burst on to the British music scene in November 1962 and over the next eight years became the most critically and commercially acclaimed band in the world. With Lennon and McCartney's short, harmonious, catchy songs, they became the best-selling band of the 20th century, and their influence on popular music was instant, lasting, and almost immeasurable. The performances of the likeable mop-topped quartet energized the music scene on both sides of the Atlantic in the 1960s and set the tone for pop music for a generation.

Formed in Liverpool in 1960, The Beatles initially derived their style from American rhythm 'n' blues music favored by Lennon's earlier band, The Quarrymen. The Beatles sharpened their sound on the club circuit in northern England and in Hamburg, Germany, and in 1961 were given a regular slot at the Cavern Club in Liverpool. It was here that they met Brian

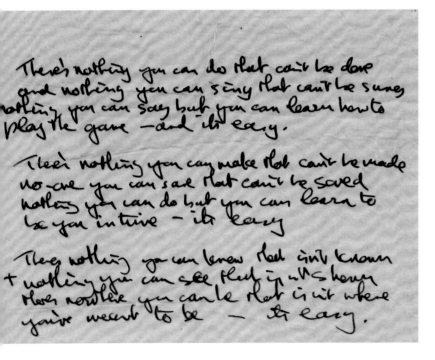

Left: John Lennon's manuscript for "All You Need is Love," the song that became a number-one hit around the world in 1967.

Epstein in December 1961, the man who became their manager and did so much to define their image and promote the band. Their first single, "Love Me Do", was released in October 1962, and was swiftly followed by a succession of chart-toppers, with "From Me To You" occupying the number one spot in Britain for 21 weeks from April 18, 1963. Beatlemania – the wild reaction of (mainly female) fans to the band – spread around the world, and when the Beatles traveled to New York in February 1964, they were met by an unprecedented crowd of 3,000 at JFK Airport. Within three months of their United States debut, they occupied the top five places in the Billboard chart. For the next two years, they played concerts in stadiums around the world, full of fans screaming so loudly that none of the band could hear themselves play. By August 1966, when

they played what became their last live concert, The Beatles had achieved eight number one singles in the USA. The band's popularity seemingly knew no bounds and it was in this period that John Lennon famously (and controversially) said "We're more popular than Jesus now."

During their career, they released 40 singles, albums and EPs that went to number one in the music charts in Britain alone; similar achievements around the world meant that their total record sales reached around one billion by 1985. The band's style evolved over the years of their collaboration, as they experimented with new sounds, instrumentation and production, and were influenced by the changing cultural developments of the mid-1960s. Their eighth album, *Sgt. Pepper's Lonely Hearts Club Band*, which was released in 1967, is still regarded as one of the high water marks of rock 'n' roll, with its skilful and imaginative blending of musical styles and studio manipulation.

With the death of Brian Epstein in 1967, the Beatles moved closer to the counter-culture of the late 1960s, and by 1968 the band began to disintegrate. They broke up acrimoniously in 1970 and never re-formed; although the individual members pursued successful solo careers, they never surpassed the defining success of their youth.

Above: October 3, 1964, The Beatles at Granville Studio, London, recording a performance for broadcast four days later for the U.S. TV show Shindig.

Below: The Beatles were constantly pursued by delirious fans: here police struggle to control over-excited girls at Buckingham Palace.

Alexander Graham Bell

1847–1922

American (from 1882)

Scientist

Alexander Graham Bell was a Scots-born scientist and inventor, whose work into the transmission of sound via electricity, led to development of the telephone, his most famous invention. He went on to found the Bell Telephone Company in the USA in 1877.

Several members of Bell's family studied speech and were elocutionists. His father, Alexander Melville Bell (1819–1905) was a teacher of elocution who published his "system of visible speech" in 1882, which showed the position of the vocal chords for each of the most common human sounds. His mother gradually lost her hearing during young Alexander's childhood, and he not only developed a sign language they could use together, but he learnt to speak clearly directly to her forehead (rather than her ears) so that she could hear him properly. Educated in Edinburgh, where he was born, as a boy Alexander was encouraged by his mother's condition to study acoustics and how sound works. Throughout his life he demonstrated a penetrating intellectual curiosity that drove him to study practical problems and create and invent devices to solve them. His family influences prompted his first interest in communication and speaking and listening devices.

In 1870, the family moved to Canada where Bell and his father continued their work with deaf people. The following year, Bell became a professor of vocal physiology at Boston University, where he experimented with a number of devices designed to transmit sound via electricity, and ultimately, "articulate speech," as he put it. He worked with Thomas A. Watson, an electrical designer and mechanic, who helped to give Bell's theories practical form. In March 1876, Bell spoke through his device to his assistant in a neighboring room: "Mr. Watson – come here – I want to see you." Three months later they extended the range to four miles and the telephone became a viable communication device.

With the telephone successfully patented, the Bell Telephone Company was founded in 1877, and by 1886 over 150,000 people in the USA owned telephones. Aged only 30, Bell could have devoted his life to marketing and refining his invention, but he went on to explore many other avenues, including the "photophone," which he called, "the greatest invention I have ever made; greater than the telephone." This remarkable device could transmit sound via a beam of light and was the forerunner of fiber optic cables. By the 1890s Bell turned his attention to the principles of flight, working with a group of young engineers to develop flying machines. The Wright brothers took off first at Kitty Hawk in 1903, but in 1909 Bell and his colleagues were responsible for the first powered aircraft to take to the skies in Canada. Bell worked with Casey Baldwin to produce a hydrofoil that set a world water speed record that remained unbroken until 1963.

Bell's scientific interests were wide ranging, and he explored heredity, invented an air conditioner, a prototype iron lung, a metal detector, the audiometer, and explored the possibility of alternative fuel sources. But the telephone is his lasting memorial, the device that made instant communication possible across continents and oceans.

Above: Bell also invented an air conditioner, a prototype iron lung, a metal detector, the audiometer, and explored alternative fuel sources.

Above: A 1910 banknote from the American Telephone and Telegraph Company, showing Alexander Graham Bell the Scottish-born American inventor of the telephone.

Below: Alexander Graham Bell demonstrating his telephone to a fascinated audience at the Lyceum Hall in Salem, Massachusetts, March 15, 1877.

Tim Berners-Lee

Born 1955

British

Computer scientist, inventor of the
World Wide Web

Tim Berners-Lee is a computer scientist who invented the World Wide Web. Berners-Lee built upon the work of many scientists before him to create the Web, and also invented HTML – the computer language with which websites, or pages on the World Wide Web, are constructed.

The Internet – in essence a global network that connects computer users to vast servers of information, first dreamt up in 1961 by engineer and scientist Leonard Kleinrock – has undoubtedly changed the world. Berners-Lee's contribution was to change how we navigated and viewed the Internet forever, by inventing the World Wide Web – an Internet-based global information medium, which is what we look at when we "browse the Internet".

Significantly, Berners-Lee made his invention free for everyone to use, rather than tying it up with legal restrictions. The World Wide Web has become a work environment in itself, a communication highway, a vast research tool, a leisure facility and a great deal more besides. It is no exaggeration to say that if the World Wide Web were to fail, the entire business world in all modern countries would collapse.

Tim Berners-Lee was born in London on June 8 1955 and went on to attend Queen's College, Oxford University in

Below: Sir Tim Berners-Lee shaking hands with Queen Elizabeth II at the re-launch of the Monarchy website, at Buckingham Palace on February 12, 2009.

1973, where he read Physics. While there he built his own computer using the pieces of an old television and a simple computer processor. He graduated in 1976 with an honors degree and immediately started working in the technology industry on barcode technology and writing software. Berners-Lee then spent 18 months as an independent consultant, which included six months at CERN (the European

Particle Physics Laboratory in Geneva, Switzerland.)

Berners-Lee wanted to enable researchers to share their work easily by storing documents and images "online" where they could be accessed at any time. In 1984 he joined CERN full time to work on the laboratory's computer system, especially the means by which computers communicate with each other. In 1989, building innovatively upon the Internet as created by his peers and predecessors, he drew up plans for a global hypertext project. Then, in October 1990 Berners-Lee designed and built the first Web server – the system to store the documents – and the first Web browser – the program to access the files. The first website was built at CERN in December and was launched on the Internet in August 1991.

Berners-Lee continued to work on designing the Web and coordinating feedback from users. He refined the specification as the World Wide Web got ever bigger and more ambitious.

In 1994 Berners-Lee founded the World Wide Web Consortium at Massachusetts Institute of Technology, with the aim of coordinating the development of the Web worldwide so that it could remain stable and reliable for users. Among many other accolades and awards, in 1999 Berners-Lee was labeled by Time magazine as "One of the 100 greatest minds of the century," and was knighted in 2004 by Queen Elizabeth II for services to the global development of the Internet.

Above: Tim Berners-Lee receiving Finland's Millennium Technology Prize for enhancing "people's quality of life" in 2004.

Left: Tim Berners-Lee speaking at the 2008 Campus Party, the world's biggest on-line electronic entertainment festival, held in Valencia, Spain.

Napoleon Bonaparte

1769–1821

French

General, statesman and emperor

Napoleon Bonaparte was the greatest French general of all time, a visionary leader and bold statesman who changed the course of French history. His great military successes throughout Europe in the early years of the 19th century restored military pride to a French nation exhausted by civil war, and when he became emperor in 1804, he introduced administrative and legal reforms, the *Code Napoléon*, that are still used in France to this day.

Born in Corsica in 1769, Napoleon was one of nine children in a middle-class family. He was educated at military school in mainland France and began his army career in 1784, and witnessed the beginnings of the French Revolution in 1789, which cast aside the nobility and the monarchy in favor of a fairer system of advancement based on hard work and merit.

Promoted to brigadier after driving the British out of Toulon in 1793, Napoleon went on to lead French troops in Italy, where he defeated the Austrians and Sardinians. Appalled by the condition of French troops, he worked to impose tighter discipline and ensure better standards for the soldiers;

Above: By 1800 Napoleon had become dictator of France and was ravenously ambitious to control and dominate Europe as a whole.

Left: Napoleon I, Emperor of the French in exile on board HMS Bellerophon, where he surrendered to the captain.

in return he won the undying loyalty of the French army. In 1798 he launched an ill-fated invasion of Egypt and his defeat by the British at the battle of the Nile prompted his return to Paris, where he launched a *coup d'état* and took control as the First Consul of France.

From 1800 he dominated French political and military life, and sought to extend his power throughout Europe. His

ambitions were viewed with alarm by other nations who fought back during the Napoleonic Wars (1803–15). Nevertheless, for a time Napoleon seemed invincible, with victories against the Austrians, Italians, and Prussians at Marengo, Ulm, Austerlitz, and Leipzig. Only the naval power of the United Kingdom prevented his invasion of Britain, and reinforced the commercial blockade of France. The costly Peninsular War of 1812–14 diverted French troops, as the British reinforced Spanish and Portuguese efforts to rid themselves of French rule.

As emperor from 1804, Napoleon reinforced his power throughout Europe by placing his relatives on the thrones of conquered nations. Napoleon swept aside many outdated French laws and introduced new legislation, the *Code Napoleon* in 1804. He restored the French economy by establishing the Banque de France, and introduced centralized educational institutions. These are a lasting legacy of his rule.

Above: French cuirassiers charging a British square during the battle of Waterloo, June 18, 1815. Almost 50,000 men were killed or wounded, and it marked the end of Napoleon's reign.

Given the scope of Napoleon's military ambitions, it is not surprising that his defeats were truly terrible, beginning with the devastating retreat from Moscow in 1812. In 1814, with many of their finest troops dead, and still more utterly exhausted, the French were almost defenseless as the British and Spanish broke out of the Iberian Peninsula and invaded France. Napoleon lost Paris and was forced into exile on the island of Elba. In one of the great comebacks of history, he escaped to rule France again during the "Hundred Days," which culminated in his ultimate defeat at the battle of Waterloo in June 1815. The emperor was exiled again, this time to the blustery south Atlantic island of St Helena, where he died in 1821.

Gaius Julius Caesar

*c.*100–44 BC

Roman

General and statesman

Julius Caesar was one the most outstanding generals of all time and an ambitious Roman statesman who made himself dictator of the late Roman republic in 48 BC. His impact on the Roman Empire was huge: he expanded the empire's boundaries, introduced political reforms and a new calendar system (the 365-day Julian calendar). The name Caesar became a title, which was used by the emperors of Germany (kaiser) and Russia (tsar). For more than 2,000 years after his death, there was at least one head of state bearing his name.

Caesar was born in 100 BC and, with the sudden death of his father in 85, he found himself the head of his notable patrician family. He joined the army, serving in Asia in 81, but returned to Rome in 78, where he was quickly embroiled in the factional world of Roman politics. He was elected to a series of posts in the *Cursus honorum*, the Roman sequence of public offices held by all ambitious politicians.

In 61–60 Caesar served as governor of the Roman province

Above: Caesar recoils in horror as he is presented with the head of his rival and former son-in-law Pompey.

Left: Caesar expanded the borders of the Roman Empire and invaded Britain in 55 and 54 BC , which was the beginning of the 400-year Roman occupation.

of Spain and on his return was elected consul for 59. Pursuing a political career in Rome was not cheap, and Caesar had incurred many debts. For this reason, the pact with his main rivals Pompey and Crassus (the richest man in Rome) was vital to his political success and their alliance, the 'First Triumvirate', was cemented by the marriage of Caesar's daughter Julia to Pompey.

Caesar was appointed governor of Gaul (modern-day France and Belgium) in 58, a post he held for eight immensely successful years, and chronicled in his writings, *De Bello Gallico*

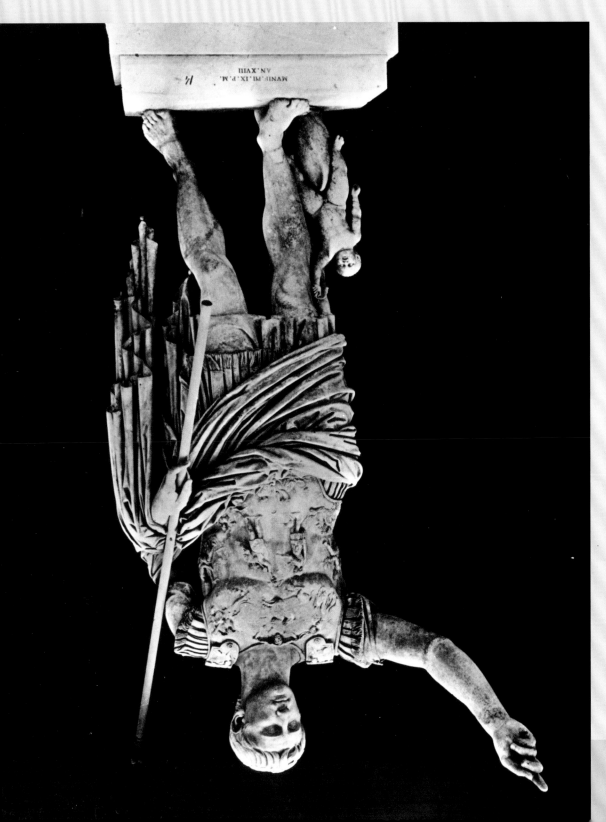

Above: Roman emperor and general Caesar Augustus, was born Gaius Julius Caesar. He held many of the great Roman offices of state, before he became dictator of the Roman Republic in 48 BC.

("The Gallic Wars"). During this time he demonstrated exceptional, if brutal, military skills as he brought Gaul under Roman control. In 55 and 54 he invaded Britain, launching a Roman occupation that was to last some 400 years.

Caesar's absence in Gaul weakened his power base in Rome. With the death of Crassus, Caesar's alliance with Pompey dissolved and when the Roman Senate threatened to try him for crimes committed during his consulship, Caesar marched his army from Cisalpine Gaul, controversially crossing the Rubicon in January 49 (the traditional border of the Roman empire in northern Italy) to fight Pompey in a civil war. Caesar defeated Pompey's forces in Italy, and was appointed dictator of Rome in 48. He pursued Pompey, who had fled to Egypt, where he was murdered. Caesar remained in Egypt, in love with the queen, Cleopatra. After beating the Pompeian generals Scipio and Cato in Africa in 46, and with a detour to Spain in 45 to crush Pompey's sons, he returned to Rome.

In a move unprecedented in Roman history, he was declared consul for ten years and dictator of Rome for life in 44. A grateful Senate named him "Father of the Country", with his person declared sacred, his image displayed on coins and statues, and the month of Quintilis renamed in his honor. But he was not universally popular among the ambitious senators of Rome, and he was assassinated on the Ides of March in 44, killed by a group of republicans led by Cassius and Brutus. Far from uniting the republicans, Caesar's death paved the way for the imperial rule of his successors under his adopted son Octavian, who became the first emperor Augustus

Charlie Chaplin

1889–1977
British
Actor and director

The most famous actor in the world from the 1920s, Charlie Chaplin was a pioneering creative influence in Hollywood as movies came to dominate a global entertainment industry. Chaplin, who appeared in over 80 films, was the most prominent face of the early movie industry in the first half of the 20th century. Widely regarded as the most accomplished comic artist of the silver screen, it is no exaggeration to call him a legend in his own lifetime.

Chaplin's life mirrored the kind of rags-to-riches tale so popular with moviegoers. Born in poverty in Victorian London, the child of music hall entertainers, he spent his early years in a series of workhouses with his brother Sidney because his mother was unable to care for them. Charlie became a professional entertainer when he launched his career in 1897 with a Lancashire clog-dancing act. In 1908 he joined the Fred Karno pantomime troupe and quickly became their star performer. Karno himself was Chaplin's first mentor, and he

never forgot his maxim: "Keep it wistful gentlemen, keep it wistful."

In 1913 the trope toured America and Charlie attracted the attention of Mack Sennett, who recruited him to work in his Keystone Comedies. Chaplin's most famous character, the Little Tramp, with his trademark baggy suit, battered Derby hat, moustache, and twirling cane, became an enormously

Above: The Kid, released in 1921 through First National, was Chaplin's first starring feature film and a huge success.

Left: A scene from The Gold Rush (1925). This is one of the most famous scenes, when the starving Tramp has to eat his boots, which were a prop made of licorice.

popular staple of 35 Keystone movies from 1915. In the world of silent film Chaplin's character was not always a tramp, but was usually down on his luck, a character on the fringes of society who represented the strength, humor and endurance of the little guy pushing against the establishment. This character more than any other seems to sum up the era of silent movies.

Chaplin's early movies were short, one-reelers which relied on physical slapstick comedy for their humor. But by 1919, together with the director D.W.Griffith, and actors Douglas Fairbanks and Mary Pickford, he decided to take more control over the whole filmmaking process by founding a new studio, United Artists. *The Kid*, released in 1921 was his first starring feature, and from 1923 to 1929, as he spent more time on production and directing, he released only three films, including, *The Gold Rush* in 1925, which has been hailed as his masterpiece.

Chaplin regarded the arrival of sound almost as a side issue. *City Lights* (1931) and *Modern Times* (1936) were essentially

silent films with Chaplin's own music and sound effects. It wasn't until 1940 that Chaplin made a "talkie," *The Great Dictator*, which viciously ridiculed Hitler and the Fascists.

Politically left leaning, and with a colorful private life, Chaplin never became an American citizen. In 1952 when he was in Britain to promote *Limelight*, his re-entry visa was revoked by the US authorities. Hurt and angered by what he regarded as a witch-hunt, Chaplin settled in Switzerland with his third wife Oona O'Neill and their growing family, returning to the USA only once in 1972 to accept a special Oscar.

Left: A very young Charlie Chaplin photographed a few years before he became the most famous actor in the world.

Above: "The Little Tramp" with his trademark baggy suit, battered Derby hat, moustache, and twirling cane, was Chaplin's most famous character.

Sir Winston Churchill

1874–1965
British
Prime minister and statesman

Winston Churchill was prime minister of Britain during the Second World War and is Britain's greatest wartime leader. He proved to be an inspirational prime minister, taking over the reigns of power in 1940, when Britain faced its most serious threat from Hitler's Germany. His rousing speeches helped to stir the country into action, boosting morale and encouraging defiance in the face of the Nazi threat. He was a diplomat and tireless administrator, who managed to secure vital support from President Roosevelt of the USA before the official entry of America into the war.

Winston Leonard Spencer Churchill was born into the British aristocracy, at Blenheim Palace, Oxfordshire, the son of Lord Randolph Churchill and his wife the American heiress Jennie Jerome. He attended the Royal Military Academy, Sandhurst, and then served with the army in India and Sudan. Next, he became a war correspondent for *The Morning Post* covering the Second Boer War in South Africa for eight eventful months in 1899.

Churchill returned to Britain and became a member of parliament

Above: Churchill in June 1941. His determination and spirit enormously bolstered the British and Allied war effort.

in the 1900 general election as a Conservative. Restless and ambitious, he was soon disillusioned and crossed the floor to join the opposing Liberal party. When they won the 1905 election, Churchill was rapidly promoted. He became first lord of the admiralty in 1911, the position he held in the early years of World War I. He took responsibility for the disaster of the Gallipoli landings in 1915 and resigned to join the army, serving on the Western Front. He returned to government in 1917 as minister of munitions, becoming secretary of state for war and secretary of state for air in 1919.

Rejoining the Conservative party, he served as chancellor of the exchequer 1924–1929, but after the Conservatives lost the 1929 election, he spent much of the next decade writing. These were years of political isolation for Churchill, as he promoted unpopular policies. He opposed the movement for Indian independence from Britain, and warned of the dangers of German rearmament, as Hitler rose to power during the 1930s.

Critical of Neville Chamberlain's government, and in particular his appeasement of Hitler, Churchill replaced Chamberlain as prime minister in May 1940, as Germany invaded France and the Netherlands. He famously told the British people, "I have nothing to offer but blood, toil, tears and sweat," and refused to negotiate an armistice with Germany, saying in a speech in June, "We shall never surrender." Crucially, he worked hard to establish

Above: Winston Churchill, surrounded by British army officers, trying out one of the new 'Sten' submachine guns early in 1941, in Kent, England.

Right: Members of Churchill's wartime cabinet on VE Day in London. Left to right, Ernest Bevin, Churchill, Sir John Anderson, Lord Woolton, and Herbert Morrison.

strong relations with the USA, realizing that Europe needed American support to win the war. He seemed tireless during the war years, working to improve British industrial output, uniting the Allied forces of the USA and the Soviet Union, and focusing completely on defeating the Nazis.

Voted out of office in 1945, Churchill was prime minister again from 1951 to 1955. He was an extraordinary man – a gifted orator, talented artist, and Nobel-winning writer. But he will always be remembered as the man who led Britain from the brink of defeat to victory in World War Two.

Cleopatra

69–30 BC

Greek
Egyptian monarch

Cleopatra is arguably the most famous queen of ancient Egypt. Popularly regarded as a femme fatale, she had a tremendous influence on the course of Roman history at a time when power was shifting among different groups. However, she was a skilled politician, who exploited Roman divisions to keep her own throne.

Originally from Macedonia in northern Greece, Cleopatra's family, the royal Ptolemy line originated with one of Alexander the Great's generals, Ptolemy I (c. 367–283 BC). The dynasty ruled over Egypt for some 300 years from 304 BC. Under the provisions of her father's will, Cleopatra was supposed to rule jointly with her brother Ptolemy XII, her brother's guardians overturned the will and ousted her from power.

Cleopatra fought back and was about to reassert her claim to the throne when Julius Caesar arrived in Egypt in 48 BC. The Roman statesman and general was actually in pursuit of Pompey, a rival for power who had fled to Egypt after a crushing defeat at the battle of Pharsalus. Pompey had been murdered, but his supporters in Egypt convinced Ptolemy to oppose Caesar. Caesar was besieged in Alexandria, but won a decisive battle on the River Nile the following year in which Ptolemy was killed. Cleopatra regained her throne, but once

Below: Through her beauty and personality Cleopatra seduced two of Rome's most powerful men, Julius Caesar and Mark Antony.

Above: Legend says Cleopatra took her own life by putting a poison asp to her breast when she heard her lover was dead.

Above right: Cleopatra ruled jointly with her brother Ptolemy XII until his guardians deposed her, but she was reinstated by Caesar in 48 BC.

again, ruled jointly with a younger brother-husband, Ptolemy XIII. The queen had a son in 46 BC and, claiming he was Caesar's, named him Caesarion in his honor.

Cleopatra followed Caesar to Rome that same year, but after his assassination in 44 BC, returned to her homeland. Rome was torn apart by civil war and two years later in 42, she met one of the men bidding for power, Marcus Antonius (Mark Antony). They embarked on a long affair, even though Antony married a wife, Octavia, in 40 BC. Cleopatra bore him twins that year and a third child was born to them in 36 BC. The pair were more than lovers – Cleopatra believed that Antony would restore the power of the Ptolemaic dynasty. In reality, his influence in Rome was ebbing away due to his prolonged stay in Egypt, and he was being openly challenged by Caesar's

nephew Octavian.

Octavian succeeded in engineering a declaration of war against Cleopatra by presenting Egypt as a threat to Rome's power. The rival forces clashes in the naval battle of Actium in 31 BC, where Antony and Cleopatra's forces were totally defeated. The two fled back to Egypt pursued by Octavian, and Cleopatra opened negotiations with him at Alexandria to save her throne. Antony was not present and, misled by a false report that his lover had killed herself, he fell on his own sword. Cleopatra soon found that Octavian was unmoved by her negotiations and she, too, committed suicide, allegedly by clasping a poisonous snake to her breast. Egypt became a province of the Roman Empire after her death.

Christopher Columbus

1451–1506

Italian

Navigator and explorer

Columbus was the greatest explorer of his age, who opened up the New World (the Americas) to European colonization. He was born in the port of Genoa in northern Italy and went to sea at the age of 14. In around 1474 Columbus developed a remarkable plan to reach India by sailing west across the Atlantic Ocean. He made several trial voyages, to the Cape Verde Islands and Sierra Leone. Satisfied that his plan was realistic, Columbus looked for wealthy patrons to sponsor a major expedition. After several rejections, he finally won the support of King Ferdinand and Queen Isabella of Spain in April 1492.

Above: On August 3, 1492, the Nina, Pinta, *and* Santa Maria *sailed westwards to catch the trade winds which blew them across the Atlantic.*

Left: When Columbus died in 1506 in Valladolid he still believed that he had reached Asia.

Columbus left on August 3, 1492 in command of 120 men aboard three small ships, the *Santa Maria*, the *Pinta* and the *Nina*. He first made for the Canary Islands and then, using the Atlantic trade winds, sailed ever more westward, despite the increasing alarm of his crews. Contemporaries believed that there was uninterrupted ocean between Western Europe and Asia, and Columbus claimed that the diameter of the Earth was much smaller than it actually is. So, when they finally reached land on 12 October, the sailors believed they had found the East Indies: in fact it was probably Watling's Island in the

Bahamas, and they named it San Salvador. Columbus explored Cuba and Hispaniola, captured some natives and then embarked on the return journey with just two ships as the *Santa Maria* had been wrecked. He anchored in home waters on March 15, 1493 and once he had displayed his extraordinary cargo of gold, spices, parrots and captives, was hailedn a hero at the Spanish court.

His second expedition began on September 25, 1493 and was a much larger affair, with 20 ships and 1,200 men to colonize the region. Land was sighted – Dominica in the West Indies – on November 3, and he pushed on. The expedition was undermined by quarrels between an increasingly dejected Columbus and his officers, and he was stuck down by a long illness while in Haiti. Columbus finally returned to Spain in 1496, convinced that Cuba was Cathay (China) and with a smaller haul of treasure. His subsequent voyages in 1498 and 1502 were marred by deteriorating relations between Columbus and his men, who complained to the king about his actions as governor. However, he did discover the South American mainland and explored the coast of Central America.

Columbus died in 1506 in Valladolid and never let go of his belief that he had reached Asia.

For centuries, Columbus was hailed as a hero, as the man who brought European civilization and religion to the New World and sent the great treasures of the Americas back home to Spain. More recently, historians have pointed out that his arrival did not have an entirely positive effect on the Americas, bringing with him, as he did, new diseases from Europe and beginning the slave trade. However, he was undoubtedly a great navigator, a tremendously brave man of great conviction, whose achievement in opening up the Americas to exploration changed the world forever.

Above: When Columbus landed in the New World, he believed he had found a new route to the East Indies. He was actually in the Bahamas.

Below: On his first voyage, Columbus sailed on to explore Cuba and Hispaniola before returning to Spain laden with captives, spices, gold, and even parrots.

Martin Cooper

Born 1928

American

Inventor and entrepreneur

Martin Cooper is generally accepted as being the inventor of the mobile phone. His role in creating and developing the first portable mobile phone has had a global impact, launching a new era in human communication.

Born in Chicago, Cooper graduated with a degree in electrical engineering at the Illinois Institute of Technology in 1950, and gained his Master's degree in 1957. After four years in the US Navy, he joined the electronics company Motorola in 1954 to develop portable products, including the first handheld police radios made for the Chicago police department in 1967.

Although Bell Laboratories had pioneered cellular communications as long ago as 1947, by the late 1960s and

Right: Released in 1983, the Motorola DynaTAC 8000X was the world's first commercially available portable cellular phone.

Below: Martin Cooper (far left, front row) with other members of the 1972-1973 Motorola DynaTAC engineering team, receiving the GlobalSpec Great Moments in Engineering award, Chicago, Illinois, October 10, 2007.

early 1970s they found themselves in competition with Motorola, using the same technology in portable devices. By then, Cooper was leading Motorola's cellular research team, and by 1973 recognized how useful it would be to carry phones around for use just about anywhere. (It is also said that Cooper was inspired by watching Captain Kirk use his "communicator" on the TV show *Star Trek* to research the possibilities of a truly mobile phone.)

After some initial testing in Washington for the Federal Communications Commission, he set up a Motorola base station in New York on the roof of the Burlington Consolidated Tower (now the Alliance Capital Building) and produced the first working cellular telephone prototype, called the

Motorola Dyna-Tac. Then, on April 3, 1973 standing on a street near the Manhattan Hilton, Cooper attempted a private call before going to a press conference upstairs in the hotel. He picked up the heavy ($2\frac{1}{2}$ lb/1 kg) handset. The phone came alive, connecting him with the base station and into the main landline system. To the bewilderment of some passers-by, he dialed the number and held the phone to his ear. Ironically, that first call made was to his rival, Joel Engel, Bell's head of research.

Cooper remembered it as follows: "As I walked down the street while talking on the phone, sophisticated New Yorkers gaped at the sight of someone actually moving around while making a phone call . . . and whilst crossing the street. Probably one of the most dangerous things I ever did in my life."

He then spent some ten years bringing the portable cell phone to the market. By 1983, Motorola was able to offer for sale a 16-oz (450-g) phone, retailing at an eye-watering $3,500 each. It took another seven years before there were a million subscribers in the United States. Today, there are more cellular subscribers than line phone subscribers in the world, with mobile phones weighing only a few ounces.

Martin Cooper went on to become chairman, CEO, and co-founder in 1992 of the wireless technology and systems company ArrayComm Inc. That call in 1973 changed the world, and it is now almost impossible to imagine life without mobile phones and the freedom they have given us.

Above: Martin Cooper at the COMPUTEX Taipei 2007 e21 Forum.

Francis Crick, James Watson, Rosalind Franklin

Francis Crick

1916–2004

British

Physicist

James Watson

Born 1928

American

Biologist

Rosalind Franklin

1920–1958

British

Chemist and crystallographer

The discovery of the building blocks of life – DNA, or deoxyribose nucleic acid – was very much a joint effort In 1953 James Watson and Francis Crick announced that they had discovered the structure of DNA, the substance that contains genetic instructions in every living thing. Rosalind Franklin's research work into molecular structure was vital to the discovery, although it was not fully acknowledged until years after her death. The work of these three scientists transformed the study of genetics.

It was not until the 20th century that the discovery of chromosomes showed what was actually responsible for giving living things their inherited characteristics. Further chemical analysis showed chromosomes to be composed of protein and DNA, and that DNA was the molecule holding the genetic information of all living organisms. Research continued to find the physical structure of the DNA molecule, and how it worked to store and transmit genetic information.

James Watson studied zoology at the University of Chicago, and graduated in 1947, before gaining his doctorate in 1950 at Indiana University. The following spring, he went to a conference in Italy, met the biologist Maurice Wilkins from Kings College, London, and saw for the first time an X-ray pattern of crystalline DNA.

Francis Crick was a physics graduate from University College, London, and later became a Cambridge PhD student, where he switched to biology and worked at the Cavendish Laboratory.

In 1951 Watson traveled from the USA to work with Crick at the Cavendish. Together, they used X-ray crystallography data (which reveals the arrangement of atoms within solid matter), to decipher the structure of the DNA molecule. Using wire structures to model the arrangement of the molecules, Watson and Crick attempted to put together their puzzle so that it would explain all the different known facts about the DNA molecule's components. Once satisfied, they published their findings in the British journal *Nature* on April 25, 1953. "This (DNA) structure has two helical chains each coiled round the same axis . . . Both chains follow right handed helices . . . the two chains run in opposite directions . . . The bases are on the inside of the helix and the phosphates on the outside . . . "

Left: Francis Crick, in Paris on April 23, 1993, explaining his work to discover the molecular structure on DNA.

Above: James Watson receiving the 1962 Nobel Prize for Chemistry, in Stockholm, Sweden.

Below: A 1956 portrait photograph of Rosalind Franklin, taken during her time at London's Kings College.

These words cleared the way for an enormous new understanding of the structure of DNA. As time and medical research has progressed, scientists have been able to work with and manipulate the information-rich DNA molecule, understanding, identifying, and curing inherited diseases in ways that were once thought impossible. Medicine and knowledge of human physiology had changed forever.

For this, Watson, Crick and Wilkins shared the 1962 Nobel Prize for Physiology and Medicine. Yet, the Nobel Committee ignored Rosalind Franklin's essential contribution. She had been seen, wrongly, as an assistant of Maurice Wilkins, though both were equally expert in showing how DNA crystallized in two different forms.

Born in London, she gained her chemistry doctorate from Cambridge in 1945 and had established herself as a world expert in the structure of graphite and other carbon compounds. Moving to London's King's College in 1951, she became an expert in X-ray crystallography, working on the structure of the DNA molecule until her premature death from cancer in 1958.

Marie Curie

1867–1934

Polish/French

Physicist

Marie Curie was a distinguished chemist and physicist, the winner of two Nobel prizes who formulated the theory of radioactivity. Her achievements were all the more remarkable, because she had to struggle against prejudices that barred women from studying at the highest levels around the turn of the 19th and 20th centuries.

Born Maria Sklodowska in Poland in 1867, Marie Curie demonstrated exceptional intellectual abilities from an early age, but was unable to pursue her education after the age of 18 because of family money troubles. She worked as a governess for two years in order to finance her sister's medical studies, on the understanding that she would do the same for her. In 1891 Marie moved to Paris where she studied physics, chemistry and mathematics at the Sorbonne, earning a degree in mathematics in 1894. She intended to return to work in Poland, but when she was denied a place at Krakow University because she was a woman, she returned to Paris, where she married fellow scientist Pierre Curie in 1895.

Pierre Curie specialized in the electrical and magnetic properties of crystals, which

Above: The dangers of radiation were not understood in the early days of research, and Marie Curie almost certainly died from its effects.

is the area of study that first drew the Curies together. Marie also worked under the guidance of the scientist Henri Becquerel, who had discovered that uranium emitted unusual rays. Curie performed a series of experiments that established that the uranium atoms were emitting an energy she termed "radiation," and used an electrometer invented by her husband to discover that the air around the uranium was a low-level conductor of electricity. This was the beginning of the Curies' work into the properties of radiation and the start of their experiments to isolate elements that were naturally radioactive. By 1898 they had discovered two new elements, which they named polonium (in honor of Marie's native land) and radium (because of its intense radioactivity). In 1903 Marie was awarded her doctorate, and in the same year, along with Becquerel and her husband Pierre, won the Nobel Prize for Physics for their joint work into spontaneous radiation.

Marie Curie's greatest achievement was to separate radium from its surrounding radioactive residues in sufficient quantities for scientist to study its properties. She also worked to stockpile radioactive elements for the purposes of research into their medicinal uses and for research into nuclear physics.

Despite the sudden death of Pierre Curie in 1906 in a traffic accident, Marie Curie continued her work and was appointed to his professorship a month after his death – the first female professor at the Sorbonne. She published the first serious article on radium in 1910 and the following year was awarded an unprecedented second Nobel Prize, this time for chemistry.

Above: Marie Curie separated radium from its surrounding radioactive residues in sufficient quantities for other scientists to study.

Below: The Curies in their laboratory in 1902. With Henri Becquerel they received the Nobel prize for physics for their work on spontaneous radiation.

Above: Marie Curie was widowed in 1906 but she continued to work and took over her husband's professorship at the Sorbonne.

For many years, Curie worked extremely hard in difficult circumstances, in ill-equipped laboratories, using substances that were to prove injurious to her health. Her notebooks remain so radioactive that they are stored in lead-lined boxes. Her fame after 1911 enabled her to encourage the French government to fund the Radium Institute at the University of Paris, and from 1922 she focused on finding practical medical applications for radioactivity.

Marie Curie died in 1934, almost certainly from an illness that resulted from her long exposure to radiation. Her discoveries were immensely important and shaped the work of subsequent scientists in splitting the atom and in using radium to treat cancers.

Leonardo Da Vinci

1452-1519

Italian

Artist, sculptor, architect, scientist, polymath

Leonardo da Vinci defines the word genius and embodied the ideal of the Renaissance man: not only was he one of the world's greatest artists, but he also possessed a penetrating intellect and an inquiring scientific mind. Although his artistic output has provided his most lasting memorial, he was also a keen observational scientist, compiling pages of anatomical and botanical drawings, scientific diagrams, architectural designs, and inventions –some 13,000 pages in total. He was, above all, an artist, and it is clear from his work that his notes were designed to explain his finely detailed drawings, rather than the other way round. Written in a cursive mirror script, surviving pages include the plans for a helicopter, a battle tank, a calculator, shoes for walking on water, and even basic theories about the Earth's movement and plate tectonics.

The illegitimate son of a peasant woman and a wealthy notary, Leonardo had little education until he was 14, when he was apprenticed to the prestigious Florentine painter Andrea del Verrocchio. Many early sketches survive from this period, showing Leonardo's early artistic skill and interest in technical matters.

In 1481 Leonardo moved to Milan where he worked for

Above: A 16th century painting of Italian artist, architect, engineer and scientist Leonardo da Vinci.

Below: A reconstruction of a swing bridge invented by Leonardo da Vinci on show at the 'The European Genius' exhibition in 2007 at the National Basilica of Koekelberg in Brussels, Belgium.

Above: The famous Codex Vaticano Urbinate 1270, also known as Leonardo da Vinci's Treatise on Painting.

the ruling Duke Ludovico Sforza and was listed in Sforza's household register as pictor et ingeniarius ducalis ("painter and engineer of the duke".) He worked on a variety of projects, including architectural plans for the dome of Milan cathedral and designs for weapons, machinery and military fortifications. The duke and the monks of Santa Maria delle Grazie jointly commissioned Leonardo to paint the refectory wall of the monastery, and the result was The Last Supper (1498), a masterful painting became one of the most influential pictures of the period.

In 1500 Leonardo returned to Florence where he worked for the notorious Cesare Borgia as chief architect and engineer. Leonardo remained in Florence until 1508, devoting much of his time to anatomical studies and to topographical surveys of the land around the city. Around 1504 he painted his most famous work, the Mona Lisa, a portrait so lifelike and enigmatic that many experts regard it as a turning point in the history of portraiture. In fact, Leonardo was so fond of the painting that he kept it in his possession at all times. In 1508 Leonardo returned to Milan, abandoning his monumental mural of the Battle of Anghiari in Florence's Palazzo Vecchio, which remained unfinished.

Between 1513 and 1516 Leonardo lived at the Vatican where he continued his studies of human anatomy but was forbidden by Pope Leo X to dissect corpses. He spent his final years in the service of King Francis I of France, and moved to Clos Lucé near the royal court at Chateau Amboise in the Loire Valley. Leonardo was reported to be a strong and handsome man and a good conversationalist with a fine singing voice. He was — most unusually for the time — a strict vegetarian. His only character flaw, it seems, was that his unflagging curiosity frequently led him to abandon one project to pursue another.

He was highly respected by his contemporaries, and his reputation as an intellectual colossus and artistic genius has remained undimmed in the 500 years since his death.

Above: The 'Mona Lisa' is unquestionably Leonardo da Vinci's most famous painting.

Louis Daguerre

1787–1851

French

Photographic pioneer and painter

Above: Daguerre helped to develop the process of using sunlight to imprint pictures on metal plates, leading eventually to the invention of the photography.

In an age when most of us can take pictures of the world around us and instantly view them on a small hand-held screen, it is hard to imagine how revolutionary Louis Daguerre's invention seemed in the middle years of the 19th century. Daguerre pioneered the first successful form of photography, the "daguerreotype," an invention which immediately attracted immense publicity: "From today painting is dead", declared the artist Paul Delaroche dramatically, and although that was far from true, the art of daguerreotypes launched a new and intriguing method of reproducing images.

Louis Daguerre was born near Paris, France and began his working life as a backdrop painter for the Opéra in Paris. In the 1820s he devised the diorama, large semi-transparent linen screens painted with images of cities or landscapes that were hung at the back of a set and skillfully lit to produce the illusion of depth and movement. Enormously popular, "diorama theaters" opened in capital cities throughout Europe. Daguerre used a camera obscura (an early type of pinhole camera) to speed the process of making sketches, and he began to experiment with ways of capturing the image of the camera obscura in order to avoid the lengthy job of tracing the image by hand.

In 1826 he met the chemist Joseph Niepce, who had been carrying out experiments using sunlight to develop images on silvered metal plates. They knew that some silver compounds were sensitive to light, so experimented with a silvered copper plate that was iodized by exposure to iodine vapor. Like many of the world's great inventions, Daguerre's breakthrough occurred by chance two years after Niepce's death in 1833. In 1835 he accidentally left a plate that had been exposed in a camera obscura in a dark cupboard for a couple of days. When he found it, he discovered that it showed a visible image, which had been developed by mercury vapor from a broken thermometer. Daguerre quickly discovered that an image could be produced after only 20 to 30 minutes' exposure, and in 1837 he found that the image could be fixed by washing the metal plate in saltwater solution.

Something of a showman, and never one to miss a chance

Above: The oldest surviving daguerreotype was produced on a silvered plate and shows a collection of plaster casts on a window edge.

Right: Daguerre's theatrical dioramas comprised large, semi-transparent linen screens, with painted pictures carefully stretched and lit to produce the illusion of reality and movement.

to profit from his work, Daguerre immediately advertised his invention, patented the process and shared the profits with Niepce's family. In January 1839, a full account of the process was given at the French Academy of Sciences, and the French government grandly announced that that the invention was free for the world to use.

Despite its immediate popularity, the daguerreotype process was flawed by modern standards: images were reversed, the equipment was bulky and the process was comparatively slow, but despite these drawbacks, the public in Europe and America flocked to have their pictures taken. Daguerre's instruction manual was published in 32 editions in eight languages, and for 20 years, until the arrival of cheaper and quicker photographic methods, the art of the daguerreotype flourished.

Charles Darwin

1809–1882

British

Naturalist

Charles Darwin's theory of evolution changed mankind's view of the natural world and humans' place in it. Darwin was a naturalist who believed that over many millions of years, all species had evolved from common ancestors by changing their characteristics in order to survive. He called this "natural selection." The impact of his truly radical thinking was extraordinary at a time when the biblical version of creation had never really been challenged. Darwin published *On the Origin of Species by Means of Natural Selection* in 1859, and most of the scientific community immediately saw the logic of his work. But many Christians were horrified, believing that Darwin's theory that man was descended from apes contradicted biblical teaching in the Book of Genesis.

In spite of this, Darwin's work on evolution by natural selection has become the unifying theory of all the life sciences. His work changed the approach not only of zoology and geology, but also botany, taxonomy, paleontology, anthropology, psychology, theology and philosophy. Even without his groundbreaking theory of evolution, Darwin would have been celebrated for his general scientific research and unorthodox thinking across a wide range of natural sciences that would

Above: A photograph of Charles Darwin taken when he was about 66 years old in 1875. His theories challenged orthodox Christian beliefs.

Left: Down House at Downe in Kent, was the home of naturalist Charles Darwin where he lived from 1859 until his death.

have made him one of the foremost figures in the history of science.

Charles Darwin was born on February 12, 1809 in Shrewsbury, Shropshire into a wealthy family. After studying at Cambridge University, he seized the opportunity to join the scientific survey ship HMS *Beagle* in 1831 on its five-year round-the-world voyage.

The *Beagle* sailed across the Pacific Ocean to South America and the Galapagos Islands. Darwin spent his time examining and recording geological features such as volcanoes, coral reefs and fossils, as well as the rich variety of flora and fauna. The extraordinary wildlife on the Galapagos particularly fascinated him, especially the finches which differed slightly, but significantly, from island to island.

The *Beagle* returned to England in 1836 loaded with

Darwin's collected specimens and extensive notes. As he wrote up his studies, worrying conclusions troubled him: all his research led to the conviction that life on Earth had continuously evolved by natural selection from simpler life forms. This evolution included humans, who had common ancestry with the apes. This was a provocative idea at a time when the received Christian interpretation of life was as the Bible said: that God had created the world and everything in it in seven days.

Reluctant to publish his revolutionary idea, Darwin quietly worked to compile cast-iron proof for his theory over the next 20 years, discussing the problems with like-minded scientific friends. It was not until another scientist, Alfred Russel Wallace, who had come to the same conclusions about evolution, announced he was ready to publish that Darwin reluctantly presented his theory, but not in person, to the Linnaean Society of London in July 1858.

His ideas changed the world of science, but it did not change Darwin. He continued working until his death in 1882, when he was buried in Westminster Abbey after a state funeral.

Above right: Tijeretas Bay in the Galapagos Archipelago. While here Darwin noticed small differences between the finches, which inexorably led him to his revolutionary theory.

Right: An original page in Darwin's handwriting from The Origin of Species by Means of Natural Selection, first published in 1859.

Below: HMS Beagle pictured in the Straits of Magellan during its circumnavigation of the globe, 1831-36. Darwin served as a naturalist on this voyage.

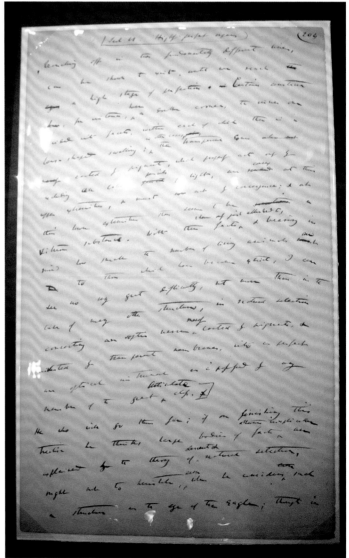

Diana, Princess of Wales

1961–1997

British

Princess

Princess Diana was one of the most recognizable figures of the late 20th century, one who, after an early and wholly unexpected death, was dubbed the "People's Princess" by the British Prime Minister, Tony Blair. Her impact on Britain, particularly the conservative world of the royal family, was striking. Popular and charming, she somehow introduced a new note of informality and accessibility into their dealings with the world, and the world, in its turn, was fascinated by her.

The princess was born Lady Diana Frances Spencer on the Sandringham estate in Norfolk, England. Diana was educated first at Riddlesworth Hall in West Heath and then at an finishing school in Switzerland. Returning to England, she took up a job as a kindergarten teacher in London. Although not especially academically gifted and somewhat shy, she performed her duties well.

Diana's life changed forever in 1980 when she was linked with Charles, Prince of Wales, the eldest son of Queen Elizabeth II. Tall, beautiful, and blonde, Diana became front-page news and, in a foretaste of what was to come, was hounded by the press, especially photographers for the tabloids. Their engagement was announced, photo opportunities and interviews were stage-managed, and in 1981 the couple were married in St Paul's Cathedral with great pomp and circumstance. Diana now took on the public role that was expected of members of the royal family. She gave birth to their first child and the second in line to the throne, William Arthur Philip Louis, in 1982 and a second son, Henry Charles Albert David, in 1984.

Diana's natural warmth and ability to relate to people from every walk of life made her extremely popular with the British public. But her relationship with Prince Charles became more and more strained in the late 1980s, and the pair finally separated in 1992. A full divorce was granted in 1996. The princess did not retire from public life, however, and apart from being a mother to her two sons, she took on an increasing number of charitable and humanitarian roles. She worked with AIDS charities, as well as those associated with children and played a prominent role in the campaign to clear abandoned landmines that were taking a heavy toll of civilian lives in many war-torn countries.

Left: Lady Diana Spencer married Charles, Prince of Wales at St Paul's Cathedral in 1981.

In 1997 Diana also appeared to have found personal happiness in her relationship with Dodi Fayed, the son of the owner of the prestigious Harrods store in central London. Rumors abounded that they were shortly to marry. However, both were killed in a car crash in a Paris underpass in 1997. Various conspiracy theories were aired after the event, but the

Above: Princess Diana, with UN Special Envoy to Angola Alioune Beye during a campaign against land mines in Angola on January 15, 1997.

Left: Princess Diana arriving with the Chairman of Christie's auction house, for a gala reception ahead of an auction of her gowns in New York to benefit AIDS and cancer charities in 1997.

most likely cause of the crash was excessive speed by their driver as he attempted to escape a pack of relentless tabloid photographers. As her brother said in his funeral address, she was "the most hunted person of the modern age." Diana's passing led to an unparalleled wave of national mourning and her funeral was watched by tens of million, not only in England but around the world.

Walt Disney

1901–1966

American

Animator, film producer and entrepreneur

Walt Disney was "the man behind the mouse", the creator of the most famous animated cartoons of the 20th century, among them Mickey Mouse and the feature-length animated film, *Snow White and the Seven Dwarfs*. He was responsible for legendary film entertainment, and, with the creation of Disneyworld, essentially created the modern theme park. Walt Disney's imagination, humor, and ability to tap into the popular taste made him a huge cultural force, as well as one of the most successful businessmen of his time.

Born in Chicago, Illinois, the young Walt Disney was a keen artist and spent a few months at the Academy of Fine Arts in Chicago before serving as an ambulance driver with the Red Cross during the First World War. On his return in 1919, Walt hoped to become a newspaper cartoonist, but instead took a job as a commercial artist with an ad agency and began to learn about animated film advertisements.

Disney moved to California in 1923 to work in Hollywood, and began to create short, animated cartoons. Mickey Mouse,

Above: Walt Disney at his drawing board in his Burbank studio. Mickey Mouse first appeared in November 1928 in the animated short Steamboat Willie.

his most famous creation, first appeared in the animated silent short, *Plane Crazy* in 1928, and later that year, after the advent of sound to the movies, acquired a voice in *Steamboat Willie*. Disney remained the voice of Mickey Mouse until 1946. Mickey was immediately popular and other characters — Donald Duck, Goofy, Pluto and Mickey's girlfriend Minnie – quickly joined him. During the Depression, the wholesome, cheery characters somehow won over a world that had little to smile about, and Disney started to make a profit. His brother Roy franchised character sales and Disney Productions began its journey to its position as one of the most powerful entertainment companies in the world.

In 1937, Disney released *Snow White and the Seven Dwarfs*, the world's first full-length animated picture, which had taken three years to produce. This was swiftly followed by *Pinocchio* (1940), *Dumbo* (1941) and *Bambi* (1942), animated movies that continue to delight audiences many years later. In the years after the Second World War, Disney captured the American mood of optimism and opportunity. Disney Studios, which was managed by his brother Roy, ventured into television, documentaries and amusement parks. The Disneys quickly saw the potential of television and produced several extremely popular series, such as *Zorro* and *Walt Disney's Wonderful World of Color*. The success of movies such as *Mary Poppins* (1964), confirmed the position of Walt Disney Productions as the most successful family entertainment empire in the world.

Disneyworld, the theme

park opened in Anaheim, California in 1955, encapsulates the Disney vision of entertainment for the family, in a colorful, clean, world that lived up to Disney's dream of a "magic kingdom." An even larger version, Disneyland, opened in Florida in 1971, followed by three more in Japan, France, and Hong Kong.

For Disney, movies were all about magic, entertainment and enjoyment – there was never any darkness. "I hate to see downbeat pictures", he said shortly before his death. The characters he created have become part of the childhoods of several generations of children, and quite simply, have given pleasure to millions.

Right: Fireworks light up the Sleeping Beauty Castle during the opening celebrations at Disneyland, Hong Kong in September 2005.

Below: In March 1954 "Uncle" Walt Disney won four Oscars at the Academy Awards, two for short films and two for documentaries.

Below right: The first Disneyland opened in California in 1955 and since then several have opened around the world. This is Euro Disney, near Paris, France, on Mickey Mouse's 75th birthday.

Bob Dylan

Born 1941

American

Musician

Above: A 20-year old Bob Dylan two years before he made his big breakthrough as a singer-songwriter.

Bob Dylan was the first – and many would say the greatest – of the singer-songwriters, whose intelligent, poetic songs encouraged and inspired other musicians. Dylan's lyrics were rooted in religion, mythology and American folklore, and the serious message underlying them was a million miles away from the frothy love songs penned by most of his contemporaries.

The combination of his distinctive nasal voice, penetrating stare and, above all, original words, ensured that Bob Dylan made an immediate impact on the music scene when he launched *The Freewheelin' Bob Dylan*, in 1963. Over 40 years later, he remains one of the most influential musicians of the 20th century, and is still acknowledged as a master by those of the 21st.

Born Robert Zimmerman in Duluth, Minnesota, and brought up in the Midwest, Dylan dropped out of the University of Minnesota, but hung around on the fringes of student life in Minneapolis, absorbing the left-leaning politics, beat poetry, and folk music, notably that of Woody Guthrie, America's great protest songwriter. He moved to New York in 1960 and began performing in the clubs and bars of Greenwich Village. The civil rights movement absorbed him, and when Peter Paul and Mary performed his song, "Blowin' in the Wind" in June 1963, Dylan came to public attention.

Claimed by the protest movement and folk singers of the 1960s as one of their own, and with his songs adopted as anthems, Dylan liked to surprise his fans. The uproar over his use of electric instruments in 1966 seems incredible 40 years on, but many of his fans, who admired the political significance of his music as much as its tunefulness, felt he had "sold out" to the commercial, disposable, world of 1960s' pop. Time has proved them wrong. Dylan has shown time and again that he is capable of writing biting lyrics and unforgettable musical arrangements. His has embraced musical styles ranging from country, blues, and rock 'n' roll, to folk and jazz.

The Times They Are a-Changin' (1964) contained more protest songs than any of his other albums, but his next, *Another Side of Bob Dylan*, reflected more personal concerns. "Like a Rollin' Stone", (1965) is often cited as his finest single recording, with its rolling chords and individual folk-rock style. *Nashville Skyline* (1969) was almost entirely a country album, while the gospel sound of *Slow Train Coming* in 1979 reflected his conversion to fundamentalist Christianity.

Despite his often sardonic manner, fans still flock to see Dylan perform live. The "Never Ending Tour" of live performances around the world began in 1988, and continues into the next century, with no sign of it stopping; they had played 2,100 shows by 2008. His songs have been covered by musicians as different as The Byrds, Jimi Hendrix, Guns 'n' Roses, and Norah Jones. His poetic, challenging lyrics have earned him several nominations for the Nobel Prize for

Above: Young protest singers Joan Baez and Bob Dylan duet during the March on Washington civil rights rally on August 28, 1963 in Washington D.C.

Left: Dylan performing for the American leg of the Live Aid concert at JFK Stadium in Philadelphia, Pennsylvania, July 13, 1985.

Below: Dylan is a musical legend, who continues to perform live, despite approaching his 70th birthday Here he sings at the 44th Annual Grammy Awards, at the Staples Center, Los Angeles in Feburary 2002.

Literature. Bob Dylan wrote some of the greatest and most memorable songs of the late 20th century and his influence on his fellow performers has been immense.

Thomas Alva Edison

1847–1931

American

Inventor and physicist

Thomas Edison was one of the most successful inventors in history, notching up a total of 1,093 patents. He devised a number of items that significantly changed the world, among them the phonograph (an early record player), the motion picture camera (the "Kinetoscope") the incandescent light bulb and an electrical power distributing system. Unlike many inventors, Edison skillfully manufactured and marketed his inventions, using the financial profits to fund his research laboratory. Edison was born into the age of steam, and by the time of his death, the world had transformed into the electrical age, largely as a result of his efforts.

Born in Milan, Ohio, Thomas Edison had little formal education and was expelled from school because his teachers thought he was retarded – in fact, he suffered from hearing problems throughout his life. He went to work as a railroad newsboy on the Grand Trunk Railroad, and during the Civil War (1861–65) worked in the telegraph section. By 1869, Edison had invented the duplex telegraph, a device enabling several messages to be transmitted at once along one wire, as well as a printer than converted the electrical signals into letters.

In 1871 Edison married and with the proceeds from the sale of the telegraph, settled in Menlo Park, near Newark, New Jersey, where he built a laboratory and machine shop. He became known as the "wizard of Menlo Park" as the world learnt of his discoveries, notably the phonograph (an early sound-reproducing machine) in 1877. Edison had been working on a modification for the telephone, and devised the carbon button transmitter that improved the clarity of sound and is still used today. He used the carbon transmitter in his tinfoil phonograph, which astonished his audience when it was unveiled in 1877.

It took another decade for him to perfect the phonograph and make it a profitable invention, however.

Above: The "wizard of Menlo Park" posing in about 1880 with the first phonograph, one of his many inventions.

Edison's increasing fame prompted investors including J. P. Morgan and the Vanderbilts to back him to form the Edison Electric Light Company in 1878. Edison did not invent the light bulb, but using a carbon filament, was the first to produce a commercially practical, long-lasting bulb, declaring, "We will make electricity so cheap that only the rich will burn candles." In 1880 Edison patented a system for the distribution for electrical power, the vital application for his light bulbs. Arguably, this is his most important invention, as it enabled the electrification of cities, and ultimately the entire world. In 1882 he switched on power distribution systems in Manhattan and in Holborn, London; by 1887 there were 121 Edison power stations in the USA.

Interested in early attempts to record motion pictures, in 1888 Edison filed a patent for a device he called a "Kinetoscope", which would "do for the eye what the phonograph does for the ear." With Kodak's development of motion picture film, Edison was able to produce a viable device and in 1896 an audience in New York City watched one of the first movies on his "Vitascope."

Edison's enormous achievements were entirely due to years of hard work and his natural intelligence – or as he famously said himself, "Genius is one percent inspiration and ninety-nine percent perspiration."

Above: Edison's Menlo Park lab after its relocation to the Henry Ford Museum in Greenfield Village, Dearborn, Michigan.

Right: Edison funded the manufacture of the "Vitascope," an early form of film projector, on condition he was credited with its invention.

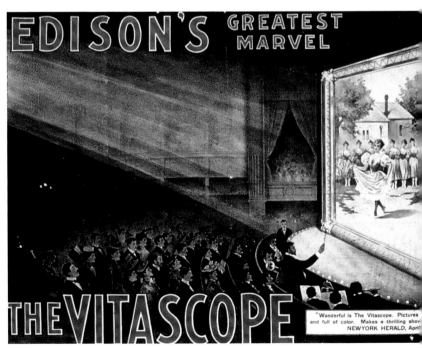

EDISON'S GREATEST MARVEL

THE VITASCOPE

"Wonderful is The Vitascope. Pictures and full of color. Makes a thrilling show. NEWYORK HERALD, April

Albert Einstein

1879–1955

German/Swiss, later American

Theoretical physicist

Albert Einstein was one of the world's greatest theoretical physicists and without doubt the most famous scientist of the 20th century. In 1905, aged only 26, he published four remarkable papers in the German journal, *Annalen der Physik*. The first was on photoelectric effects showing how light interacts with matter. The second was about Brownian motion and provided direct evidence of molecular action, all of which became part of atomic theory. The third and most important paper discussed electrodynamics and introduced

Einstein's radical theory of special relativity. Simplified down to $E = mc^2$, this work led to the development of nuclear power. The fourth paper used his special relativity equations to show that miniscule amounts of mass could be converted into vast amounts of energy — a vital step in the later development of nuclear power.

Albert Einstein was born into a Jewish family in Ulm, Württemberg, Germany on in 1879. Even as a child he stood out as a mathematician, and by the time he was 12 he had learned Euclidean geometry. He later trained to become a teacher of physics and mathematics, but was unable to get a teaching post, and instead became assistant examiner of patents for electromagnetic devices at the Swiss Patent Office. It was here in his spare time that he came up with his world-changing theories.

Following the publication of his astounding papers in 1905, Einstein quickly moved up the academic ladder. In 1909 he became professor extraordinary at Zurich; in 1911 professor of theoretical physics at Prague; in 1912 professor of theoretical physics at Zurich; in 1914 director of the Kaiser Wilhelm Physical Institute and professor at the University of Berlin.

At this time Einstein believed that the special theory of relativity must work in tandem with the theory of gravitation and this led to the publication of his paper on the general theory of relativity in 1916. Not content with this, Einstein also investigated the theory of radiation. In 1921, he was awarded the Nobel Prize for Physics for "his services to Theoretical Physics, and especially for his discovery of the law of the photoelectric effect."

By the early 1930s, with the rise of the Nazi party and increasing anti-Semitism, Germany was an uncomfortable

Below: Albert Einstein delivering one of his popular recorded radio lectures for American broadcaster NBC in 1955.

Above: Einstein with the astronomer Dr J. A. Miller discussing the periodic table at Spoul de Swartmore Observatory, c.1925.

Right: Theoretical physicist Albert Einstein, in December 1943, when he was living and working in the USA after fleeing Nazi Germany.

place, for a highly regarded Jewish physicist. Einstein renounced his citizenship and emigrated to the United States in 1932, where he became professor of theoretical physics at Princeton University and an American citizen in 1940.

Still intellectually questioning, Einstein wanted to solve several more problems of physics. He tried to resolve Newtonian mechanics with the laws of the electromagnetic field and studied the problems created by quantum theory and the movement of molecules. In another area, he investigated the thermal properties of light with a low radiation density and here his observations led to the beginnings of the photon theory of light.

Einstein's research into the nature of matter paved the way for other scientists to create nuclear power and the atomic bomb. Physicists continue to build on his work, discovering things such as gravitation waves, that he predicted.

Michael Faraday

1791–1867

British

Physicist and chemist

Faraday was a physicist and chemist who made several discoveries that are now fundamental to science. He put together two laws of electrolysis and three laws of electromagnetic induction, and is generally considered to be the greatest experimental physicist of all time. Faraday created the first electrical generator and the first transformer, both devices that control the flow of electricity. He greatly expanded human scientific knowledge and it is largely due to his research that electricity became the world's most widely used source of power.

The son of a blacksmith, Faraday was born in humble circumstances in south London. He was apprenticed to a bookbinder and in 1812, having become fascinated by some of

Above: Faraday's large horseshoe electromagnet dates from around 1830. His discovery of electromagnetic induction led to the development of modern generators.

Left: Faraday holding up the bar of his heavy glass. He is generally considered to be the greatest experimental physicist of all time.

the scientific books he was binding, he applied to the famous chemist Sir Humphrey Davy (1778–1829) for a position as his assistant, having attended one of his lectures at the Royal Institution. Davy took him on, and Faraday gained his scientific education at the feet of the man who did so much to transform chemical knowledge in the early years of the 19th century.

By 1816 Faraday was conducting his own research and made important contributions to analytical chemistry. He also invented an early form of the Bunsen burner. In 1827 he succeeded to Davy's chair of chemistry at the Royal Institution.

From 1830 Faraday increasingly turned his attention to physics and electricity and what became his great life's work, *Experimental Researches on Electricity*, which was published over the course of the next 40 years. He reported on the fundamental laws of electrolysis in 1832 and 1833, with what became Faraday's Laws. He experimented with electricity and magnetism, discovered electromagnetic induction, and used this principle to construct the electric dynamo, the forerunner of modern generators. In 1845 he worked to show that the forces of electricity, magnetism, light and gravity are connected. He also invented the transformer to control the strength of an electrical current. He was the first to use electrolysis, employing electricity to break down chemical compounds.

Faraday had no students, very little social life and only one longsuffering assistant, yet he was reportedly a charming man whose annual Christmas lectures at the Royal Institution began a tradition that continues to this day. Much of his scientific research was immensely complicated, and yet he had the ability to make science accessible to a wide audience. In addition to his scientific research, Faraday carried out valuable public service, working for both the government and private business. He investigated the causes of explosions in coalmines, was consulted on air pollution in London, and in 1855 write to *The Times* on the subject of sewage in the River Thames. He was consulted on the construction of lighthouses, on planning the Great Exhibition of 1851, and advised the National Gallery on methods of preserving the national art collection.

Faraday's reputation extended beyond the scientific world and he was held in high esteem. Offered a knighthood by Queen Victoria, he declined it, although he gratefully accepted a small cottage at Hampton Court Palace in 1858.

Below: Faraday lecturing at the Royal Institution in December 1855 in front of Prince Albert. Faraday was very keen to teach science to a wide audience, and the annual lectures continue to this day.

Alexander Fleming

1881–1955

Scottish

Bacteriologist

Alexander Fleming discovered penicillin in 1928, the antibiotic drug that transformed the treatment of bacterial infections. Before the use of penicillin, people died from simple infections such as a sore throat, or from infected wounds. A drug that could be manufactured synthetically, cheaply and quickly, penicillin was the first antibiotic to be used in medicine, and its commercial use from the mid-1940s quite simply saved millions of lives.

Educated in Scotland, Fleming became a shipping clerk before embarking on a career in medicine. In 1902 he won

Right: Fleming holding a Petri dish similar to the one used when he noticed that penicillium notatum *destroyed staphylococci bacterium.*

Below: Alexander Fleming being decorated with the Legion of Honour at the Academy of Medicine in Paris, by General Charles de Gaulle.

a scholarship to study medicine at St Mary's Hospital Paddington, London, and in the years immediately after qualifying, he assisted the pioneering bacteriologist Almroth Wright, who developed a vaccine for typhoid fever. During the First World War, Fleming and Wright served with the Royal Army Medical Corps, where they made great advances into the use of antiseptics in dressing wounds. After the war Fleming was appointed director of the department of systematic bacteriology at St Mary's and made a number of important discoveries. The first, in 1921, was the existence of the enzyme lysozyme in nasal mucus, tears, and saliva, a naturally occurring antibacterial enzyme that protects the body against infection. Fleming's study of lysozyme was an important contribution to knowledge about how the body fights infection.

Fleming's most important discovery occurred by accident in 1928. He noticed that mold was growing in a Petri dish containing staphylococci bacterium, and that the mold was destroying the bacteria with which it came into contact. Fleming made a culture of the mold and realized that it could have great biological significance. He identified it as penicillium notatum, a type of mold that also grows on stale bread, and carried out experiments that proved it could destroy certain bacteria. His 1929 paper, "On the antibacterial action of cultures of a penicillium, with special reference to their use in the isolation of B. influenzae", has been called one of the most important medical research papers ever written, but although Fleming could explain the properties of penicillin and envisage its potential, he lacked the chemical knowledge to isolate and stabilize it for use as a drug.

The true significance of Fleming's discovery was not recognized until 1940, when Howard Florey and Ernst Chain succeeded in purifying and stabilizing penicillin, realized the great therapeutic effects of the drug, and began work on large-scale production in the USA in 1941. The timing was perfect, given the fact that the world was at war. It has been estimated

Above: Scottish bacteriologist Professor Alexander Fleming photographed in November 1943, soon after penicillin was first commercially manufactured.

that the use of penicillin from 1942 saved an additional 12–15 per cent of lives among servicemen suffering from infected wounds. It was also used to treat bacterial diseases, such as pneumonia, scarlet fever, gonorrhea, diphtheria and meningitis, and scientists went on to develop further antibiotics such as ampecillin and tetracycline.

In 1945, Fleming, along with Florey and Chain, was awarded the Nobel Prize for Physiology or Medicine for "for the discovery of penicillin and its curative effect in various infectious diseases."

Benjamin Franklin

1706–1790

American

Statesman and scientist

Benjamin Franklin, one of the first great Americans, had wide-ranging interests and a superb mind. Printer, scientist, diplomat, publisher, politician and writer, he was one of the Founding Fathers of the American republic and helped draft the Declaration of Independence. He dabbled so successfully in scientific experimentation that he made fundamental contributions to the science of electricity, and was also an accomplished and witty writer.

Above: Franklin and his son William using a kite and key during a storm to prove that lightning was electrical, June 1752.

Left: Benjamin Franklin in about 1750, during the period when he was investigating and observing the effects of electricity.

Franklin grew up in Boston, but moved to Philadelphia as a young man. Apprenticed to his older brother as a printer, he published *Poor Richard's Almanac* from 1732, a bestselling directory of household hints, Franklin's wisdom, and seasonal weather forecasts. Published for 25 years, the Almanac gave Franklin financial security, which was reinforced by other well-paid printing commissions, including that of producing the paper money for the state of Pennsylvania. By 1748, Franklin was wealthy enough to retire from trade and lead the life of a gentleman engaging, as he put it, in "Philosophical Studies and Amusements."

Franklin's research into electricity was far more than an amusement, however, and resulted in his election to the prestigious Royal Society in London. He published his findings as *Experiments and Observations on Electricity* in 1751, which was translated into German, French and Italian by the end of the century. Franklin's famous experiment with the kite in a thunderstorm helped establish the link between lightening and electricity, but equally important, he invented a battery to store an electrical charge and invented the vocabulary of the electrical world: conductor, charge, discharge and electrify. His other inventions ranged from the lightening rod to bifocal spectacles.

His work as a statesman probably had the most immediate impact on the events of the 18th century. Active in Pennsylvania politics from the 1730s, Franklin was dispatched to England in 1757, where he remained with only a two-year break (1762–64) for the next 18 years. His mission was to insist upon the rights of the colonists to tax their own people and to contest the rights of the British authorities to gather taxes from the American colonists without parliamentary representation. Franklin believed in the British Empire and wrote over 100 articles between 1765 and 1775 attempting to guide the opposing sides to agreement. But in 1775 on his return to America, Franklin was an active participant in drawing up the Declaration of Independence, the document that launched the Revolutionary War (1775–1781), noting that the colonial leaders needed to remain united, "Yes, we must, indeed, all hang together, or most assuredly we shall all hang separately." Franklin's diplomatic skills were employed once more in 1776, when he went to Paris to negotiate for practical military support for the colonists; the military intervention of France was eventually critical for the American victory in 1781.

In 1783 Franklin became American ambassador to Paris and on his return to Philadelphia he was elected president of the state of Pennsylvania, a post to which he was twice re-elected. A participant at the Philadelphia Convention that drafted the US constitution, Benjamin Franklin was undoubtedly the most famous American of the 18th century and one of the most influential in that country's history.

Below: The title page of the fifth edition of Experiments and Observations on Electricity, *published in London in 1774. By this time, Franklin was back in America, using his diplomatic skills to prevent war between Britain and the American colonies.*

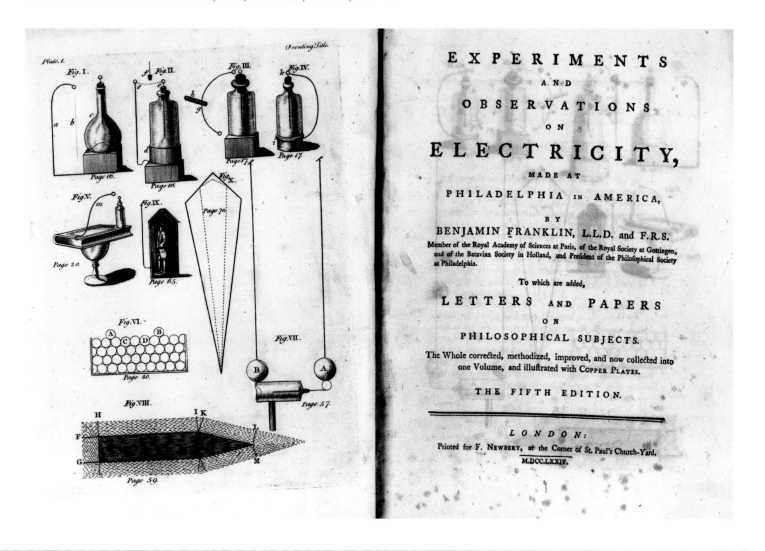

Sigmund Freud

1856–1939
Austrian
Psychiatrist

Sigmund Freud was an Austrian psychiatrist who founded the study of psychoanalysis. He developed important and influential theories about the functioning of the mind, and tried to show that the unconscious mind is essential to understanding conscious thought and behavior.

Sigismund Freud was born in Frieberg, Moravia on May 6, 1856. The oldest of eight children, Sigmund moved to Vienna with his family in 1860 and lived and worked here for the majority of his life. He displayed a keen intellect at an early age, and although his parents were poor they struggled to ensure he received a good education. Freud graduated with honors from secondary school and, after abandoning his plans to study law, enrolled in the Medical College of the University of Vienna in 1873.

Freud graduated from medical school in 1881 and begrudgingly accepted a job as a doctor at Vienna General Hospital, although he had little interest in standard medicine. In 1896 Freud married Martha Bernays; they would go on to produce six children, the youngest of whom, Anna, would continue her father's work in psychiatry and become a noted doctor herself.

Freud began experimenting with hypnosis to treat his neurotic patients, but he quickly disregarded that method for an exercise first introduced by Dr. Josef Breuer that would eventually be known as the "talking cure." Freud thought that his patients' problems may be rooted in disturbing or traumatic episodes from their past, and encouraged them to speak freely about their lives. He discovered that this often led to a reduction or even an end to their neurotic behavior.

After writing *Studies in Hysteria* with Breuer in 1895, Freud began to focus solely on psychoanalysis. His next book, *The Interpretation of Dreams*, published in 1900, brought Freud's theories to worldwide acclaim and is generally considered to be his most influential work. He followed that success with *Psychopathology in Everyday Life* in 1901 and *Three Essays on the Theory of Sexuality* in 1905, which has endured as one of the most controversial volumes of his career. It caused him to fall out of favor among his colleagues, many of whom believed he placed too much emphasis on the psychosexual element of personality development. He gained newfound respect when he gave a speech to the International Psychoanalytical Congress in Salzburg in 1908 and his fame grew even further after traveling to the United States, where he gave a celebrated series of lectures that would

Above: Freud in his late 70s. His concepts of the id, ego and superego are widely accepted and used by professional psychoanalysts.

Above: Freud with his daughter Mrs Hollistschek and the Welsh psychoanalyst Dr Ernest Jones, who was Freud's official biographer, photographed in about 1938.

Right: Sigmund Freud's study photographed in 1905. When he moved to London in 1938 he brought most of the furnishings with him.

ultimately be published as *Five Lectures on Psycho-Analysis*.

The legacy of Sigmund Freud continues to be a subject for debate among scholars. In 2001 *Time* magazine listed him as one of the 100 most influential thinkers of the 20th century, while five years later *Newsweek* magazine labeled him "history's most debunked doctor." Although some of Freud's pioneering theories are now regarded with skepticism, his ideas about conflict between instinct (the *id*) and conformity (the *superego*) within our minds have become widely accepted. His presence is still felt today, – the term "Freudian Slip" is an example of how his influence has permeated popular culture. Despite being a polarizing figure, he is largely recognized as the undisputed father of modern psychiatry.

Galileo Galilei

1564–1642

Italian

Astronomer and scientist

The famous physicist Steven Hawking has said, "Galileo, perhaps more than any other single person, was responsible for the birth of modern science." Galileo made important contributions to the sciences of motion and astronomy, and established the principle of using the language of mathematics to study the natural world. His belief in the Copernican view

Above: Galileo published studies of gravity and motion and in 1589 became professor of mathematics at the University of Pisa.

Left: When he heard about the invention of the telescope in the Netherlands, Galileo made his own, in 1609.

of the universe (in which the planets revolve around the Sun, rather than the Earth) was a revolutionary step that earned angered the Catholic Church.

Galileo was born in Pisa, Italy in 1564. His father, a musician, had always intended for his eldest son to study medicine. At the age of 11 Galileo was sent to study in a Jesuit monastery where he quickly became fascinated with the traditions of religious life and, after a few years, declared his desire to become a monk. This did not please his father, who sent him to the University of Pisa to study medicine when he was 17. Galileo, however, became fascinated with mathematics, and left university in 1585 without a degree, supporting himself

for the next few years by teaching mathematics.

It was during this time that he began his studies on motion, famously observing a swinging chandelier in the roof of Pisa cathedral. Using his heartbeat to keep time, he discovered that the amount of time between each swing was the same regardless of the arc of the swing. This would become known as the "Law of the Pendulum" and would eventually be used to regulate clocks.

Galileo's reputation improved as he published a number of papers on gravity and motion. In 1589 he became professor of mathematics at the University of Pisa and challenged one of Aristotle's laws of physics – that larger, heavier objects fall at a faster rate than smaller, lighter ones. Galileo conducted several experiments, the most famous being from the top of the Leaning Tower of Pisa, and proved conclusively that Aristotle had been wrong.

In 1609, after learning of the invention of the telescope in the Netherlands, Galileo built his own and began to study the heavens. He discovered that exterior of the moon is rough, (it had previously been assumed to be a smooth surface), saw that four moons revolve around Jupiter, and observed spots

Above: Galileo was religious, but his announcement that the Earth revolved around the Sun contradicted Church teachings. He was tried for heresy and found guilty, but then quietly continued his scientific research at home.

on the Sun. His study of astronomy led him to question the universally accepted theory that the Sun rotated around the Earth. He advocated the heliocentric theory, first put forward by the astronomer Nicholaus Copernicus nearly a century earlier, that that the Earth and all heavenly bodies orbited the Sun.

Galileo was a religious man, and while he believed in the Bible, he publicly questioned its interpretation. His views were not popular with the Church and after standing trial for heresy in 1616 (for which he was acquitted), he was forbidden to teach. Galileo remained silent for several years, but in 1630 published *Dialogue on the Two Great Systems of the World, Ptolemaic and Copernican*, which discussed the controversial theory. In 1633 the Inquisition found him guilty of heresy. Galileo, by this time blind in one eye, continued his scientific research at home until he died in 1642 with a tarnished name and reputation.

Indira Gandhi

1917–1984

Indian

Politician and prime minister

Indira Gandhi was elected prime minister of India from 1966 to 1977, and again from 1980 until her assassination in 1984. She was India's first female prime minster. Politically left-leaning, she forged an alliance with the Soviet Union, led India to victory in the 1971 conflict with Pakistan, and

established India as a nuclear power.

Gandhi was only child of Jawaharlal Nehru, the leader of the Indian independence movement and, from 1947, the country's first prime minister. She attended Santiniketan University and Somerville College, Oxford before marrying Feroze Gandhi in 1942. Shortly after their wedding they were jailed for 13 months for taking part in a protest against British Rule in India. Both she and her husband would remain active in politics, with Feroze Gandhi winning a seat in the Indian parliament, while Indira became a personal advisor to her father. Feroze died in 1960, leaving her to raise two sons, Rajiv and Sanjay.

When her father died in 1964, Mrs. Gandhi was persuaded to take up a career in politics herself. She was elected as a member of parliament in her father's Indian National Congress Party, and was appointed a minister in the cabinet of Prime Minister Lal Bahadur Shastri. When Shastri died in office in 1966, Gandhi campaigned successfully to replace him as party leader, and thus prime minister of India. As prime minister, Indira shrewdly used every political tool at her disposal to consolidate her power and authority.

In 1971, a hugely popular Gandhi won re-election but found herself charged with election fraud. In 1975 a high court upheld the charges and Gandhi lost her seat in parliament, as well as the premiership. To prevent this, she introduced a state of emergency that lasted for over two years. During this period she implemented several unpopular policies, including sterilization as a means of widespread birth control. By 1977, when she initiated new elections, she severely misjudged public reaction to her extreme actions and was voted out of office.

Gandhi returned to office as prime minister in 1980, amid turmoil in India as several states tried to gain independence from central government. Sikh extremists in Punjab used violence, and in 1984 an armed Sikh freedom fighter group barricaded itself inside the Golden Temple, Sikhism's holiest shrine. Despite the large number of civilians present, Gandhi ordered the army to invade, and several hundred people were killed. Five months later, Indira Gandhi was assassinated by two of her own Sikh bodyguards on October 31, 1984 as retaliation for her actions.

Left: Indira Gandhi became the first female prime minister of India in 1974 and established India as a nuclear power.

Above: Gandhi addresses a crowd of students at New Delhi on December 1971, the year she won re-election but found herself charged with election fraud.

Right: Indira Gandhi with her father Jawaharlal Nehru and US President John F. Kennedy and his wife Jackie in 1961.

During her four terms as prime minister, Indira Gandhi is credited with leading India to victory in the war for the liberation of Bangladesh in 1971. She signed the Indo-Soviet Treaty of Peace, Friendship and Cooperation in 1971. She strengthened India's economy by nationalizing the banks. Gandhi was a champion of the "Green Revolution," an agricultural initiative to ensure that India could feed itself. She was a strong leader, who reveled in the power of high office. Many regard her as one of the greatest Indian prime ministers.

Mohandas "Mahatma" Gandhi

1869–1948

Indian

Political activist and spiritual leader of India

Mahatma Gandhi was one of the most important figures in the campaign for Indian independence from the British Empire. The figurehead of the nationalist movement from the 1920s, he achieved his aims by using *satyagraha*, or passive resistance. In reply, the British authorities repeatedly imprisoned Gandhi, in 1922, 1930, 1933 and 1942, hoping to break his spirit. They failed, and Gandhi is upheld today as the model of what can be achieved by peaceful protest.

Mohandas Karamchand Gandhi was born in Porbander in present-day Gujarat, western India. Gandhi traveled to London to study law at University College in 1888, with the intention of becoming a barrister. He was called to the bar at the Inner Temple and returned to Mumbai (then Bombay) to start his own practice, but it failed. Instead, in 1893 he accepted a one-year contract to work in Natal, South Africa for an Indian law firm. He was horrified by the active and often violent discrimination against Indians, and this awakened in him a burning desire to fight racism and social injustice. He

Below: Gandhi and his followers during the Salt March of 1930, just one of Gandhi's acts of non-cooperation against the British colonial rulers.

founded the Natal Indian Congress in 1894 to unite the Indian community into a political force, and organized a mass protest, using for the first time his still-evolving philosophy of *satyagraha*. The policy of non-resistance was ignored by the authorities, which attacked, jailed, and even shot protesters. But their brutality eventually backfired, and an embarrassed South African government had to negotiate a compromise.

In 1915 Gandhi returned to India and joined the Indian National Congress Party that campaigned for freedom from British rule. By 1921 Gandhi was its spiritual leader and encouraged the boycott of British- and foreign-made goods to revitalize local businesses and economies. He preached non-violent passive resistance and started a campaign of mass civil disobedience. He gave up wearing Western clothes, adopting instead simple homespun garments, and began weaving his own cloth, as did his followers. His policy of non-cooperation with the British included strikes, a refusal to pay taxes and a refusal to respect colonial law. The famous 240-mile salt march to Dandi in 1930 to by-pass a salt tax was a typical example of his campaigning. He commanded huge influence and respect across India, and around this time he was given the name *Mahatma*, meaning "Great Soul."

Gandhi led the "Quit India" movement during the Second World War, which was ruthlessly suppressed by the British. Independence, when it finally came in 1947, did not run smoothly. Although the handover was peaceful, civil war between Hindus and Muslims appeared inevitable, as the suBcontinent split into the Muslim Pakistan and East Pakistan (now Bangladesh), and the largely Hindu India. Vast numbers of people were dispossessed and left with no option but to travel hundreds of miles to safety in their new homeland. Spontaneous fighting broke out across the country and almost a million people were slaughtered.

Gandhi worked with the British authorities to sort out the enormous problems and began his last fast-unto-death, in an

Above: Outside 10 Downing Street, London, where Gandhi attended the 1931 Round Table Conference on Indian constitutional reform.

effort to stop the violence. On January 20, 1948 an attempt was made on Gandhi's life. He survived, but ten days later, on the evening of the 30th, a Hindu fanatic assassinated him in the gardens of Birla House in New Delhi.

A man of great moral courage, Gandhi won the admiration of millions of Indians, as well as many Europeans. He is still upheld as the finest example of peaceful but forceful protest by people across the globe.

Bill Gates

Born 1955

American

Computer programmer and entrepreneur/
philanthropist

Bill Gates is one of the richest men in the world and founder of Microsoft, the world's largest personal computer and software company. His work in developing the MS-DOS computer operating system and then licensing it to IBM in 1981 has led to Microsoft's dominance of the world computer market.

While still at the Lakeside School in Seattle, Washington, Gates started his first company with lifelong collaborator, Paul Allen. It was called Traf-O-Data, and analyzed traffic patterns. After the success of that initial venture, the school hired them to develop a program for scheduling school classes. Gates used his new invention to place himself in classes with mostly female students. He graduated from the Lakeside School in 1973, and after scoring 1590 out of a possible 1600 on the Scholastic Aptitude Test, set off for Harvard University.

Gates spent only two years at Harvard before dropping out to take advantage of a burgeoning new industry. After reading an article in *Popular Mechanics* magazine about the introduction of a new microcomputer called the "Altair", Gates immediately saw the need for programming new computer software. In 1975 Gates and Allen formed Microsoft and started work on developing programs for new computers being manufactured by the fledgling Commodore and Apple companies. Gates' big break came in 1980 when Microsoft was hired by International Business Machines, (IBM), to format a new operating system for their new personal computer. Gates constructed the Microsoft Disk Operating System, (MS-DOS), and it was at this time that he displayed the first glimpse of his keen business savvy by selling the licenses to the new software to other developers. MS-DOS became the standard software on which all new personal computers would run and hundreds of millions of copies were sold.

In 1981 Gates continued Microsoft's expansion by crafting a new operating system, originally entitled "Interface Manager", but later known to the world by the name of "Windows." He also developed a new "web browser," a device used by computers

Above: Bill Gates at the launch of Windows 98 in San Francisco, California. His company Microsoft dominates the world computer market.

to interact with the information, text and pictures on the Internet, called "Internet Explorer." He shrewdly combined this new device with the existing Windows software and the result was an instant smash hit. It is estimated that Windows, or at least one of its many components, which include Windows Office Manager and Windows Media Player, have been used on over 90% of the world's personal computers.

Microsoft is the world's biggest and most successful software company. Gates has brushed off serious charges in

Above: At Shoreline Community College, near Seattle, President Bill Clinton and Bill Gates participate in a roundtable discussion on job retraining. Gates' enormous wealth means that he has access to world leaders.

Right: Having made his fortune, Gates and his wife founded The Bill and Melinda Gates Foundation to donate money to an extensive list of charities. It is the largest privately owned charitable foundation in the world.

the USA that accused Microsoft of having illegal monopoly on the computer market. The case was ultimately settled out of court.

Bill Gates has used his enormous wealth to establish, along with his wife, The Bill and Melinda Gates Foundation. They have donated money to an extensive list of charities, including global health initiatives, education, and agricultural projects around the world. The Foundation has made a special project of equipping public libraries in the United States with computers. Gates has said that he wants to ensure "that anyone who can reach a library can reach the internet."

Bob Geldof

Born 1954

Ireland

Musician and political activist

As a young man lacking ambition, Bob Geldof bounced from job to job, and at one point found himself squatting in an unused building in London. By 1986 he had transformed himself into a respected musician and one of the foremost charity donors in the music industry. In a rather unlikely turn of events, since 1984 this Irish rock star has worked tirelessly to relieve poverty and famine in some of the poorest nations on Earth.

Geldof began his musical career as a rock journalist in Canada, writing for a small, independent newspaper called *Georgia Strait*. It was his first encounter with success: he enjoyed writing about music, his articles proved popular with readers and he quickly became a recognizable face in Vancouver. His status as an illegal alien in Canada forced him to return to Dublin where, at the urging of friends, he formed a band called Nightlife Thugs, which was later changed to the Boomtown Rats.

While the Boomtown Rats would enjoy modest success, it was in 1984 that Geldof found himself horrified and distracted by a news report about the ongoing famine in Ethiopia. Unable to sit idly by, Geldof hatched a plan for raising money. His idea was a charity single recorded by the world's biggest music stars with all proceeds going to the beleaguered nation. The project would be known as Band Aid.

Geldof worked tirelessly to ensure that everyone involved, including the recording company of Phonogram Records, the performing artists, and the retailers selling the disc, would receive no profits. Geldof teamed up with Midge Ure to co-write the single, "Do They Know It's Christmas," and several

Below: British musicians including George Michael, Bono, McCartney, Freddie Mercury, and Bowie gather on stage at the finale of Live Aid, the concert Geldof organized to raise money for the Ethiopian famine.

of the most famous names in music, including Sting, Bono, George Michael and Phil Collins, among many others, lined up to support the cause and record the song. Geldof had modest hopes for the effort, but the recording rocketed to Number One and eventually earned over £5 million.

Delighted by the success of Band Aid, in 1985 Geldof organized a benefit concert event dubbed Live Aid. The main stages for this multi-venue, simultaneous concert were Wembley Stadium in London and JFK Stadium in Philadelphia. Other countries joined in with their own shows, and over 400 million people viewed the event as it was broadcast around the world. It was the single largest satellite hook-up telecast in broadcasting history at that time, and raised approximately £150 million (around $300 million). Twenty years later Geldof repeated the feat, with the Live 8 concerts in 2005, held as part of the UK's "Make Poverty History" campaign.

Above: Geldof speaking about social responsibility and conscience at a seminar for business leaders in Helsinki, Finland, October 2, 2007.

Left: Bob Geldof at Wembley Stadium, London, looking forward to his hugely ambitious Live Aid concert, July 10, 1985. An estimated 400 million people watched the show in 60 countries across the world.

Following pages: The crowds gather for the Live Aid concert on July 13, 1985 in London, England. Two Live Aid concerts were held at the same time, one at Wembley Stadium, London and the other at JFK Stadium, Philadelphia.

He was knighted by Queen Elizabeth II for his contributions toward eliminating poverty and world hunger. A long list of awards would soon follow, including two nominations for the Nobel Peace Prize in 2006 and 2008. Despite near universal praise for his charitable works, Geldof humbly shrugs off his nickname, "Saint Bob." As he told *Rolling Stone* magazine, "I don't want to be Saint Bob because halos are heavy and they rust very easily, and I know I have feet of clay because my socks stink."

Ivan Getting

1912–2003

American

Physicist and electrical engineer

Where am I? The question seems simple enough, but in historical terms, the answer has never been easy. For centuries, navigators and explorers had used the stars to determine where they were, and humankind has continued to search the heavens, looking for a way of fixing more accurately their position on the globe. Ivan Getting is credited with devising the Global Positioning System, or GPS, the device that makes traditional maps redundant and enables precise and efficient navigation around the world.

Ivan A. Getting was co-inventor of the GPS and had actually dreamed up the basic idea in the 1950s, while serving as vice-president of research and engineering at the Raytheon Corporation. Born in New York City, he earned his physics degree as an Edison scholar at the Massachusetts Institute of Technology (MIT) in 1933. He attended Oxford University as a graduate Rhodes scholar and earned his DPhil in astrophysics in 1935.

Above: Ivan Getting in his offices at the Aerospace Corporation.

Left: Julian Hartt (left) and Ivan Getting (right) inspecting a model of the Mercury-Atlas launch vehicle in 1961.

Above: A modern, civilian, GPS unit on display at the CeBIT trade fair in Hanover on March 4, 2008.

During most of his career, Ivan Getting directed his scientific, technical and administrative energies to the support of the US military effort. He was responsible for developing a number of missile systems during the 1940s and 1950s, including the weapons system that successfully shot down the German V-1 rockets that targeted London towards the end of World War II.

Getting's idea for a global positioning system was relatively simple. He would use the electronic signals from satellites on fixed orbits around Earth to provide positioning data that could be received by computer systems on Earth. While serving as the founding president of Aerospace Corps from 1960 to 1977, he advanced his idea of using a system of satellite transmitters and atomic clocks to allow the calculation of precise positioning data for rapidly moving vehicles ranging from cars to missiles. Getting's proposal was quickly adopted. When the US Air Force launched their final Navstar satellite into orbit in 1995, they completed a network of 24 satellites known as the Global Positioning System – the GPS. Although GPS was initially developed for the US military to guide missiles to targets, it is now routinely used for air traffic control systems, ships, trucks and cars, mechanized farming, search-and-rescue, tracking environmental changes, and more.

Getting was awarded the 1989 Institute of Electrical and Electronics Engineers (IEEE) Founders Medal for, "leadership of critical programs and enterprises in radar; advanced electronics, space and navigation as well as service to the engineering profession." As a result of his vision and dedicated work, anyone with a relatively inexpensive GPS receiver can instantly learn their location on our planet, including latitude, longitude, and even altitude, to within a few hundred feet. This incredible technology was made possible by a combination of scientific and engineering advances, and by co-operation from a variety of government departments.

That Ivan Getting played a very major part in all of this is without doubt. It is thanks to him and a team of dedicated scientists that we need never again be lost – as a hiker, sailor, explorer or mere wanderer.

Mikhail Gorbachev

Born 1931

Russian

President of the Soviet Union

Above: Gorbachev at the Congress of Deputies, Moscow, the day before he took the oath as executive President of the USSR.

The first Soviet head of state to be born after the Russian Revolution of 1917, Gorbachev was also the leader who dismantled the Soviet Union, and brought Russia's Communist era to an end. In the process, he oversaw the move towards democratic reform in Eastern Europe, the demolition of the Berlin Wall, and the ending of the Cold War. Thanks to Gorbachev, decades of division between Russia and the democratic countries of the West were ended, and the world would never be the same.

Born into a peasant family in the steppes of southern Russia, Mikhail Sergeyevich Gorbachev was a boy of 10 when the Germans invaded in World War II, and the suffering they caused made a lasting impression on him. He went on to study law in Moscow, where he met and married his wife Raisa. He also joined the Communist Party, and for the next 25 years he rose through its ranks. In 1979 Gorbachev joined the Politburo (the USSR's ruling body), and in March 1985 he became its general secretary, and the leader of the Soviet Union.

He immediately set about reforming the country and economy through *glasnost* (openness) and *perestroika* (re-

Left: Geldof receiving the Man of Peace Award from former Soviet president Mikhail Gorbachev and Major of Rome Walter Veltroni.

structuring). Despite opposition, he encouraged greater economic ties with the West, and widespread political reform at home. The Chernobyl nuclear disaster in 1986 helped to underline the need for modernization and the dangers of the country's ageing nuclear arsenal. Gorbachev had already addressed this problem by negotiating nuclear arms reduction treaties with US President Reagan.

Other immense changes were to follow. In 1988 Gorbachev withdrew Soviet troops from Afghanistan. Then he abandoned the long-standing Brezhnev doctrine – the Soviet control of

Above: Gorbachev and Reagan watch as Secretary of State George Schultz and his Soviet counterpart Eduard Shevardnadze sign a declaration of intent. Gorbavhev and Reagan made huge strides in dismantling the world's nuclear arsenal.

East European affairs. From then on, the countries of the Eastern bloc could determine their own national policies.

Throughout 1989 a series of largely peaceful revolutions led to democratic reform across Eastern Europe, and while Communist hard-liners opposed Gorbachev's reforms, the world applauded his actions. In 1990 he was awarded the Nobel Peace Prize for effectively bringing the Cold War to an end. In the Soviet Union, growing food shortages highlighted the country's economic problems. *Glasnost* had also encouraged the resurgence of national identity, and increasingly, Soviet republics such as the Ukraine, the Baltic states and Azerbaijan, declared their wish to leave the Soviet Union. Nationalism was also on the rise in Russia, and in the aftermath of a failed nationalist coup, the Soviet Union collapsed, as its constituent states declared their independence. Gorbachev's attempts to

forge a new democratic union were thwarted, and he stood down on Christmas Day, 1991. The Soviet Union was officially dissolved the following day, ending 74 years of Communist rule. Boris Yeltsin became the president of the newly created new Russian Federation.

Since then, Gorbachev has been honored as one of the greatest men of his age. *Time* magazine named him one of the most influential people of the 20th century, and while the reforms he began have not necessarily followed the path he would have liked, few will argue that thanks to him, the world is now a better and safer place.

Ernesto "Che" Guevara

1928–1967

Argentinean

Revolutionary

During his lifetime "Che" Guevara, the Marxist revolutionary, political leader and freedom fighter, caught the public imagination through his revolutionary achievements in Cuba. After his death he became an international symbol of political

idealism and revolutionary fervor. His portrait by Alberto Korda was once dubbed the most famous photograph in the world, and transformed Che into a pop culture icon.

As a young man, the Argentinean-born medical student traveled through Latin America on a motorcycle, and he was struck by the poverty, inequality and exploitation he witnessed. As a result he realized that only revolution would improve the lot of the Latin American people. He met Fidel Castro in Mexico, and joined his 26th July Movement, dedicated to the overthrow of the US-backed Cuban dictatorship of Fulgencio Batista. In December 1956 they embarked on a campaign of rebellion in Cuba. Guevara soon rose to prominence as a guerrilla leader, and as the revolutionary movement flourished, he became Castro's most trusted lieutenant.

On New Year's Day 1959 Batista fled the country, and when Castro assumed power, Guevara played his part by convicting the Batista supporters accused of war crimes, and exiling Castro's political opponents. After a spell as Cuba's minster of industry, Guevara embarked on a worldwide diplomatic tour. In the process he became an international figure, with the popular appeal of a charismatic rock star. Unlike many political figures of his day, his interviews revealed a straightforwardness and honesty which helped to reinforce his political message, and which appealed directly to the young.

In December 1964 he traveled to New York to address the United Nations, where he gave a passionate speech, protesting against apartheid in South Africa, segregation in the United States, the US embargo of Cuba, and the exploitation of the third world. In the process he demonstrated his transformation from a guerrilla leader into a world revolutionary. Che was also a prolific writer. His manual on guerrilla warfare became essential revolutionary reading, while his memoir of his motorcycle journey though South America received international praise.

Then, in February 1965 he dropped out of the political limelight. He had become increasingly convinced that his true vocation was as a guerrilla leader. He traveled to the Congo, where he offered his support to the revolutionary movement. He soon became disillusioned with the rebels, who, he claimed lacked real political will. Turning his back on the Congo, he

Left: Guevara briefly became Cuba's Minister of Industry after Castro assumed power before going on a world diplomatic tour. Charismatic and charming, he was a popular figure.

Above: Guevara attended the United Nations Trade Conference at the Palace des Nations, Geneva to protest about the US embargo against Cuba.

Right: Che Guevara watches as Fidel Castro lights his cigar during their guerrilla campaign in the Sierra Maestra mountains of Cuba, 1956.

next surfaced in Bolivia, where he commanded a small force of guerrillas. However, in October 1967 he was cornered and captured by Bolivian troops, and on October 9 he was executed by his captors.

Nelson Mandela said of Che that he was, "an inspiration for every human being that loves freedom." Others see him as a terrorist, a ruthless advocate of a failed ideology, or as simply a romantic in camouflage gear. *Time* magazine named him one of the 100 most influential figures of the 20th century, and 40 years after his death, Korda's famous image of Che has become a global brand.

Johannes Gutenberg

c.1398–1468

German

Inventor and printer

In 1450 Johannes Gutenberg put together his first printing press in Strasbourg and in the process revolutionized the world. Before his invention, every single written word was hand-written and then painstakingly hand-copied to create a single, precious book. In the 14th century woodblock printing presses already existed, but the process was laborious and inflexible. With Gutenberg's printing press of movable type, books could be produced quickly, in numbers, and at a much more affordable price.

Above: Sadly, thanks to financial differences with his partner, Guttenberg probably lived the last few years of his life in penury.

Left: Guttenberg checking a page in his workshop while printing his first 42-line Bible on his movable metal type printing press.

Gutenberg had to overcome a number of technical difficulties before his printing press could work. He did not so much invent it, as put together and improve a number of existing technologies to make the printing process a viable business. Many historians regard Guttenberg's printing press as the single most important invention of the second millennium.

The development of the printing press meant that the Bible — the most significant book of the time — could be printed in large numbers and distributed widely for everyone to read, especially after it was translated from Latin into the vernacular language of each area. The creation of mass printed books allowed for the spread of scientific and cultural ideas, and opened the way for the European Renaissance. It also helped the spread of Protestantism during the 16th century.

Johannes Gutenberg was born in about 1398 in Mainz, Germany into a wealthy merchant family. Few details are known of his early life, although he probably trained as a goldsmith, which provided the metallurgical skills essential for the development of movable type. At some stage he moved to Strasbourg.

Gutenberg worked hard to resolve his ideas and probably paid out a good deal of money on perfecting his press, as he certainly got into debt. He had to be able to produce individual letters (type) in sufficient numbers for printing purposes. He also had to design the type and the molds to make it, as well as develop an alloy soft enough to cast easily, but hard enough to take the pressure of the heavy press. Furthermore, he had to develop an oil-based sticky ink that would work with the type. He also needed to find an economic source of paper, rather than expensive calf-skin vellum. The printing press itself was developed from similar screw-type wine presses used in the Rhine valley.

In about 1452 Gutenberg started work on his first Bible, which was completed in an edition of between 180 and 200 copies in 1455, most on paper but some on vellum. It had 1,280 pages, laid out in two 40-line columns per page (this changed to 42 lines for later editions). Unfortunately, Gutenberg and his financial partner Johann Fust had a dispute over money, and despite the success of the Bible, Gutenberg was effectively bankrupt. He seems to have managed to start up a small printing shop in Bamberg, but as none of his printed books bear his name or a date, it is difficult to be certain.

Gutenberg's press transformed the world by enabling the spread of information and ideas far more quickly and easily than at any time in the past.

Below: Many historians agree that Guttenberg's printing press is the single most important invention of the entire second millennium.

Sir Edmund Hillary

1919–2008

New Zealand

Mountaineer

On May 29, 1953 Edmund Hillary, succeeded in a monumental quest where many others before him had failed, becoming the first man to ascend to the summit of Mount Everest. Accompanied by Tensing Norgay, a Nepalese Sherpa and local mountaineer, Hillary set foot atop the world's highest mountain a little over 24 hours after the last phase of the expedition had begun.

Hillary took an interest in mountaineering during a school trip to Mount Ruapehu as a teenager, and went on to climb many of New Zealand's tallest peaks. Sir John Hunt recruited him for the British Everest Expedition, which was sponsored by the Joint Himalayan Committee of the Alpine Club of

Above: Prince Charles, Sir Edmund Hillary, and Prime Minister Helen Clark at the Aotea Centre on March 9, 2005 in Auckland, New Zealand.

Left: Tensing Norgay known as "Sherpa" Tensing, and Edmund Hillary enjoying a snack on their return from the summit.

Great Britain and the Royal Geographic Society. The news of his accomplishment reached England the night before the coronation of Queen Elizabeth II and played a significant role in bolstering the morale of a country still recovering from the war.

Hillary and Tensing spent only 15 minutes at the summit of the mountain, where they took pictures and left a small cross as a memorial, but that short visit would be the impetus for many of Hillary's future achievements. He went on to lead the New Zealand contingent of the Commonwealth Trans-Antarctic Expedition of 1955–1958, and became the first man to reach the South Pole in a motor vehicle. He continued his exploration of the peaks in the Himalayas for much of the 1960s, but eventually turned his attention to the people of Nepal. He established the Himalayan Trust, which built the Khumjung School in 1961 and the Kunde Hospital in 1966.

Hillary's conquest of Mount Everest led to a surge in the numbers of tourists to the area around the base of the mountain and the Nepalese people cut down vast swaths of the nearby forests in order to provide for the influx of sightseers and mountaineers. Concerned about the impact on the environment in the wake of this new, booming tourism, Hillary persuaded the New Zealand government to assist Nepal in establishing the area around Mount Everest as a National Park. Laws were set in place to protect the local forests and Hillary dedicated the remainder of his life to championing various environmental causes. He was the Honorary President of the American Himalayan Foundation, a charity to preserve the ecology of the Himalayas, and Honorary President of Mountain Wilderness a group dedicated to the preservation of mountains around the world.

Hillary was knighted for his achievement in June 1953, and to mark the 50th anniversary of his accomplishment the Nepalese government granted Hillary honorary citizenship, the first foreign national to receive that honor. Immediately after his famous climb he published his first book, in association with Sir John Hunt, entitled *The Ascent of Everest*. A second book, *From the Ocean to the Sky*, was a detailed account of his exploration of the River Ganges from its mouth to its source in the Himalayas. Hillary died at the age of 88 in 2008 in his home country of New Zealand.

Above: Edmund Hillary at base camp on Mount Everest, in 1953. With Tensing Norgay, he became the first man to conquer Everest.

Right: Sir Edmund Hillary arriving for the Mount Everest Golden Jubilee Celebration at Durbar Square in downton Kathmandu in May 2003.

Alfred Hitchcock

1899–1980
British
Film Director

Alfred Hitchcock is the most famous filmmaker to hold the dubious distinction of never winning an Academy Award, despite being nominated in the Best Director category five times. The film director, who specialized in psychological thrillers, and became known as "the master of suspense", began his career in 1921 when he bulldozed his way through the doors of the Famous Players-Lasky Company in London to demonstrate his skills as a "title card" designer. His

Above: Janet Leigh, Hitchcock's murdered blonde in his 1960 thriller Psycho, speaks at the launch of a stamp with Hitchcock's portrait.

Left: Known as "the master of suspense," Hitchcock specialized in psychological thrillers and liked to make a fleeting cameo appearance in each of his movies.

professionalism earned him a solid reputation and by 1923 he had moved on to Gainsborough Pictures Studios, where he was promoted to art director for the film *Woman to Woman*. Later that year, when the director of the film *Always Tell Your Wife* became too ill to complete it, Hitchcock stepped in to finish the picture. His bosses were so impressed that he was hired to direct a series of movies, mostly romantic melodramas, which eventually led to what is now considered the first Hitchcock movie, *The Lodger*, in 1926.

Hitchcock toiled for several years as a director of mediocre silent movies, before making a move to British International Pictures. He teamed up with Michael Balcon of Gaumant-British Pictures and, after a dismal start, he produced a string of films that are considered his best of the British years, including *The 39 Steps, Sabotage* and *The Lady Vanishes. The Man Who Knew Too Much*, from 1934 was so popular that Hitchcock remade the film in 1956 to even greater acclaim.

Hitchcock could no longer resist the urge to work in Hollywood, and in 1940 made his first American movie, *Rebecca* for David O. Selznick. Four of Hitchcock's films would compete for the Best Film Oscar during his career: *Suspicion, Foreign Correspondent* and *Spellbound*, but only *Rebecca* would win the trophy in 1940. A series of memorable, classic films followed, including *Strangers on a Train, Rear Window, Dial M for Murder, The Trouble with Harry, Vertigo, North by Northwest* and *Psycho*. Hitchcock was nominated for the prestigious Directors' Guild of America prize for each of those films, eight

times in fact, and, again, he never won. But the movie industry was not alone in snubbing the world's most famous director – Alfred Hitchcock was nominated for two Emmy awards for directing episodes of his hugely successful television series, *Alfred Hitchcock Presents*, and he lost both of those in 1956 and 1959.

Despite the bewildering lack of peer recognition, Alfred Hitchcock has earned an enduring place in movie history, with 53 feature films to his credit. Many modern-day directors including Steven Spielberg, Quentin Tarantino, John Carpenter and Brian De Palma have spoken of Hitchcock as an influence on their work. Part of the fun of watching a Hitchcock movie is trying to find the cameo appearance of the director himself; he made a habit of appearing, in some form, in every picture he directed. In 1967 the Academy of Motion Pictures Arts and Sciences, honored Alfred Hitchcock with the Irving Thalberg Memorial Award at which time Hitchcock gave the shortest acceptance speech in Academy Award history by simply saying, "Thank You."

Below: Hitchcock filming The Man Who Knew Too Much, *a 1955 Paramount remake of his 1934 spy thriller.*

Adolf Hitler

1889–1945

Austrian

Dictator

Not all those who changed the world did so in a positive way. As the leader of the fascist National Socialist German Workers' Party (NSDAP) – the Nazis – Adolf Hitler blamed the interwar woes of Germany on Europe's Jewish population, and on those of the political left. The result was the Second World War and the Holocaust. As Germany's *Führer* (head of state), his policies eventually plunged the world into a new global conflict. It ended only after his death, and the devastation and defeat of his German Reich. In the process, as many as 72 million of the world's population lost their lives, the majority of whom were civilians. Since Hitler's death, mankind has struggled to recover from the horrors he unleashed, and to ensure that such evil can never be allowed to happen again.

After the First World War, Hitler embraced anti-Semitic, anti-Communist and nationalist views, and he found a voice in the Nazi party. By 1921 he became the party leader, a post he held throughout his imprisonment, in the wake of a failed coup attempt. It was in prison that he wrote *Mein Kampf* (My Struggle), which laid out his extremist ideology.

After his release in 1924 he set about winning control of Germany through the ballot box. During the next decade the percentage of the Nazi vote climbed steadily, and in January 1933 Hitler became the German Chancellor. Within a year he managed to establish totalitarian rule in Germany. For the next five years he consolidated his grip on the country through propaganda, the raising of German prestige, the re-armament of the military, and the repression of the "enemies" of the Third Reich.

State-sponsored persecution of Jews, gypsies and political opponents gathered pace during the later 1930s, and while some of Hitler's domestic policies might have been beneficial, such as the building of the autobahns and the introduction of the affordable Volkswagen motorcar, much of this was a by-product of his expansion of Germany's military power base.

After the annexation and capture of Austria, Czechoslovakia, and the Rhineland, Hitler invaded Poland in 1939, which lead to war with France and Britain. German conquests in Poland, France, the Balkans and the Low Countries led Hitler to expect further easy victories, and so in July 1941 he invaded the Soviet Union. Although German tanks reached the outskirts of Moscow, the Russians drove the invaders back, and by April 1945 their troops reached Berlin. The United States entered the war in December 1941, and by 1945 Allied troops were invading Germany from east and west. With the Third Reich in ruins, and Russian troops approaching his Berlin command bunker, Hitler took his own life.

It was only then that the world realized the full scale of the horror Hitler had inflicted. Over six million Jews had been murdered, together with eight million other victims of the Nazi death camps. Since then, the victims of the Holocaust have been remembered, while the world has tried to expunge the last traces of the man who ordered their death. If Hitler made a positive contribution to world history, it was that his actions led to the creation the United Nations, in an attempt to avoid any repetition of the Hitler years.

Above: Hitler rose to power as the leader of the fascist National Socialist German Workers' Party, and became Chancellor of Germany in 1933.

Above: With his determination to expand Germany and to rid the country of "non-Aryan" people, Hitler is directly responsible the Second World War and the deaths of some 72 million people .

Below: Hitler addressing the Reichstag in May 1941. By all accounts he was a compelling and charismatic speaker.

Edwin Hubble

1889–1953

American

Astronomer

Edwin Powell Hubble was an American astronomer who changed forever our understanding of the nature of the Universe by showing that there were other galaxies besides the Milky Way. He also demonstrated that the universe is expanding, when it was long thought to be static.

Edwin Powell Hubble was born on November 20, 1889 in Marshfield, Missouri but moved in the same year to Wheaton,

Above: From 1919 until his death in 1953 Hubble worked on the staff at Mount Wilson Observatory, near Pasadena, California. He used the Hooker telescope to observe that there was more to the Universe than just the Milk Way

Left: Edwin Hubble beside the 18-inch Schmidt telescope at Caltech-owned Palomar Observatory in California, in 1949.

Illinois. At the University of Chicago he concentrated on mathematics, astronomy, and philosophy, graduating in 1910. He spent the next three years in the UK as one of Oxford's first Rhodes Scholars, originally reading law, before switching to Spanish and gaining a master's degree.

He gained his astronomy doctorate in 1917 at the Yerjes Observatory of the University of Chicago, with a thesis on the

photographic investigation of faint nebulae (galaxies). By 1919 he was offered a staff position at the famous Mount Wilson Observatory, near Pasadena, California.

At that time, astronomers believed that the whole universe consisted entirely of the Milky Way. But at Mount Wilson, Hubble was able to use the new 100-inch (2.5-meter) Hooker Telescope – the world's largest at the time – to prove conclusively that the nebulae he observed were much too distant to be part of the Milky Way, and were actually entire galaxies outside our own. Although many in the astronomy establishment opposed this idea, Hubble announced his discovery on January 1, 1925 and changed once and for all our view of the universe.

Hubble also used the Hooker telescope to measure the movement (also known as redshift) of some galaxies away from each other. He could then estimate a value for the rate of the universe's expansion. This became known as Hubble's Law in 1929. In turn, this discovery was the first true evidence for the "Big Bang" theory of the creation of the universe. It is the most important development in our understanding of the cosmos since our early ancestors first looked up at

Above: The Hubble Space Telescope is named after one of the greatest modern astronomers who changed peoples' understanding of the universe.

the stars and wondered about it all.

Having discovered the existence of new galaxies, Hubble went on to devise the most commonly used system of classifying them. He arranged different groups according to their shape and appearance in what became known as the Hubble Sequence.

Hubble remained on the Mount Wilson staff until his death on September 28, 1953 from a blood clot in his brain. His successors have revised some of Hubble's calculations to estimate that the Universe has been expanding at the same pace for between 10 and 20 billion years.

Hubble's name lives on in the names of schools, colleges, and, most famously, the orbiting Hubble telescope. Launched in 1990 (and repaired by astronauts from the Space Shuttle in 2009) the Hubble telescope remains in orbit around the Earth and enables astronomers to view space without the distortion produced by the Earth's atmosphere.

Steve Jobs

1955-2011

American

Computer technologist and entrepreneur

The introduction of cheap personal computers in the last decade of the 20th century changed the lifestyles and working practices of the entire world. No single name has achieved more distinction and status than Steve Jobs' company Apple. This firm has positioned itself at the top end of the computing market and continues to lead the industry in innovation with its award-winning Macintosh computers, iPod music players, and software.

Apple is a global influence in the digital music revolution, having sold almost 200 million iPods and over six billion songs from its iTunes online store. To complete the picture, Apple also entered the mobile phone market with its revolutionary iPhone.

Jobs also co-founded the Pixar Animation Studios, which has created eight of the most successful animated films of all time, including *Toy Story*, *A Bug's Life* and *Wall-E*, winning 20 Academy Awards and making more than $4 billion at the worldwide box office to date.

Born in 1955 in Los Altos, California, Jobs grew up in the apricot orchards that later became known as Silicon Valley. After high school, he studied briefly at Reed College, Oregon and then went on to work for the Atari Corporation in 1974. Jobs renewed his acquaintance with Steve Wozniak an old friend from an earlier summer job, saying Wozniak "was the first person I met who knew more about electronics than I did."

They went to meetings of the local amateur Homebrew Computer Club and worked to design computer games for Atari. With additional funding, Jobs and Wozniak launched their own business from the Jobs' family garage in 1976. They succeeded in their first commercial venture when the Byte Shop in Mountain View, California bought their first 50 fully assembled computers in the same year. The name for their corporation was based on Jobs' favorite fruit, and the logo was chosen to play on both the company name and the word "byte."

The genius of Steve Jobs was to recognize the need to make computers user-friendly, and to that end he devised an intuitive graphical user interface (GUI) operated by a keyboard and mouse pointing device. In 1984 the release of the first Macintosh sparked a revolution in the computing world: instead of a blinking cursor and lines of code, the screen displayed images, icons and responsive boxes. Furthermore, the screen and hard disk were all housed in one unit. It was a seductive, well-engineered and popular alternative, but Apple struggled to compete with Microsoft, which dominated the computing market by providing the operating software for far cheaper mass-market machines.

Through the early 1980s Jobs controlled the business side of the corporation, successively hiring presidents who would take the organization to a higher level and becoming a multimillionaire before the age of 30. Despite boardroom battles in the late 1980s and 1990s, Job has driven Apple to produce innovative and groundbreaking products, from the iMac to the iPod and iPhone, which have changed communication and entertainment across the world.

Above: Jobs co-founded Apple Corporation with his old friend Steve Wozniak, someone, he said, who knew more about electronics than he did.

Above: In 1997 Apple was in financial trouble, but Jobs worked hard to publicize their technology and turned the company around.

Below: Jobs is the dynamite behind the success of Apple, ensuring the company is always at the leading edge of accessible technology.

Right: Apple CEO Steve Jobs discussing a new version of the Mac-Book during a launch at Apple Headquarters October 14, 2008, in Cupertino, California.

John Paul II

1920–2005

Polish

Pope

Polish-born Karol Józef Wojtyla, was elected pope on October 16, 1978, and succeeded Pope John Paul I, who died suddenly after only a month in office. He became the first non-Italian pontiff in over 450 years and by the time of his death almost 27 years later, had enjoyed one of the longest papal reigns in history. Widely acclaimed as one of the 20th century's most influential leaders, John Paul II was one the most traveled in world history. He made many journeys, visiting over 100 countries worldwide. He was also a champion of human rights, despite maintaining somewhat conservative views on social issues such as abortion, homosexuality and contraception.

In 1979, a year after being elected, John Paul II returned to visit his homeland, including his birthplace of Wodice. Millions turned out to see him; but rather than criticizing Poland's Communist government, he simply gave the Polish people a sense of hope. To many, he was an inspiration and a nationalist figurehead, who was a major influence in the rise of Solidarity, the Communist bloc's first free trade union. Its leader Lech Walesa recalled how the Pope brought back moral values: "They defeated the Communist system . . . the Pope is the author of the victory over Communism."

The pope's meeting with Russian leader Mikhail Gorbachev at the end of 1989 – the "year of revolutions" in

Below: Karol Józef Wojtyla became Pope John Paul II when he was elected by the College of Cardinals on October 22, 1978 at Saint Peter's Basilica in the Vatican City.

Eastern Europe – was an historic encounter between the man who nudged the Communist bloc towards freedom and the man who let them go. It set the seal on a decade of change and collapse of Communist regimes. Without the Polish pope it would all have taken much longer.

John Paul II had a deep commitment to the developing world and was an outspoken advocate for a better deal for the poor. Yet, he had no time for those in the Church who proposed radical political solutions to such great problems. In Latin America, he condemned the 'popular church' of left-wing priests, and despite the enormous Catholic populations in Central and South America, he seemed more at ease during visits to Africa – particularly South Africa after the end of apartheid.

In Asia, the most remarkable event of his many visits was his celebration of mass in the Philippine capital Manila. One million people had been expected – but four million attended, making it the biggest papal mass in history.

However, of all his journeys, that in 2000 to the Middle

East was the culmination of a dream. Retracing the steps of Jesus, he traveled through the minefield of the region's politics, encouraging reconciliation between Israel, the Palestinians and the three great faiths.

By 1998, his 20th year as Pope, he suffered increasingly poor health, with the slurred speech and trembling hands of Parkinson's disease. Despite this he continued his papal duties until his death in his Vatican apartments on April 2, 2005. John Paul II's last words before his death were, in his native Polish, "Let me go to the house of the Father."

Revered in his lifetime, since his death there have been widespread calls for his canonization as a saint.

Above: Pope John Paul II is surrounded by pilgrims as he visits the shrine of the Virgin Mary on August 14, 2004, in Lourdes, France.

Left: Polish Cardinal Wojtyla became Bishop of Krakow in 1958, and then archbishop in 1963, before ascending to the papacy, age 58, to become pope.

John Fitzgerald Kennedy

1917–1963

American

35th President of the USA

John Fitzgerald Kennedy was only in office for a thousand days, but in that time he came to symbolize the hopes of the Western world for a brighter future. The idealism of the time was epitomized by the creation of the Peace Corps, his commitment to civil rights and his promotion of America's space program. He summed up his vision with inspirational words in his inaugural address, "And so my fellow Americans, ask not what your country can do for you; ask what you can do for your country." Moreover, at a time when the Communist forces of the Soviet Union apparently threatened the West, Kennedy faced them down; the specter of Communism retreated back to Eastern Europe and the arms race slowed down.

Kennedy was born in Brookline, Massachusetts into a high profile, political, Irish-American family. He graduated from Harvard in 1940 and entered the U.S. Navy. Kennedy became a national hero in August 1943 when PT-109, the torpedo

Above: Kennedy was the first Catholic and the youngest person, to be elected by the Democratic party to stand for and become president of the USA.

Left: A powerful and inspirational orator, Kennedy worked to establish civil rights and set in motion legislation to eliminate segregation.

boat he commanded, was rammed and sunk by a Japanese destroyer. Despite serious back injuries (which troubled him for the remainder of his life), Lieutenant Kennedy towed a badly injured member of his crew back to his base, receiving the Navy and Marine Corps Medal for his heroism.

In 1946 Kennedy became a Democratic congressman representing the Boston area, and then in 1953 a senator – the same year that he married the glamorous Jacqueline Bouvier. All his life Kennedy battled illness of one sort and another, most seriously, aged only 30, he was diagnosed with Addison's disease, which was kept a close family secret. Senator for Massachusetts from 1952–1960, Kennedy won the Democratic nomination to run for president in 1960, announcing, "We stand on the edge of a New Frontier." His TV debates with

Republican rival Richard Nixon were watched by millions, many of whom saw the clean-cut young senator as their hope for the future rather than the saturnine Nixon. Kennedy promised to revive the American economy and eliminate poverty, and won the presidential race by a very narrow margin — some speculate that his notorious father Joseph Kennedy "bought" a number of critical votes.

Kennedy intended to establish civil rights and set in motion legislation to eliminate segregation. In 1961 he committed the USA to land a man on the Moon by the end of the decade. Unknown to most, he also continued and escalated President Eisenhower's involvement in the conflict in Vietnam.

In 1960 Kennedy supported a group of Cuban exiles that intended to invade their homeland and overthrow the Communist government of Fidel Castro. It was a disaster culminating in the failed invasion at the Bay of Pigs. Two years later, Cuba was the center of action again during the Cuban Missile Crisis. Russian premier Khrushchev finally backed down and agreed to remove missile launchers and nuclear weapons deployed on the island: the crisis was resolved, world war averted and both sides stepped away from the brink. In 1963 they agreed a test ban treaty and the nuclear arms race slowed.

Kennedy was assassinated at the height of his popularity in Dallas on November 22, 1963, and the circumstances of his death remain a subject for conspiracy theorists.

Above: Kennedy's inauguration ball: From the right, President-elect John Fitzgerald Kennedy, First Lady Jacqueline Kennedy, and Vice-President Lyndon Johnson.

Below: The President and Chancellor Konrad Adenauer surrounded by a crowd at Cologne Cathedral, during Kennedy's visit to Germany in June 1963.

Jack Kilby, Robert Noyce

Jack Kilby
1923–2005
American
Physicist

Robert Noyce
1927–1990
American
Inventor and entrepreneur

The invention of the integrated circuit – the microchip – laid the foundation for the entire field of modern microelectronics that now feature as everyday items in all our lives. From those first simple circuits has grown a worldwide integrated circuit market whose sales in 2007 alone totaled $219 billion. It is the legacy of two remarkable men.

It was only a transistor and other components on a slice of the element germanium that Jack Kilby showed to a small group of his colleagues at the Texas Instruments (TI) semiconductor laboratory in Dallas on September 12, 1958. None there could have predicted that they were seeing in

Above: Jack Kilby was the first to patent the integrated circuit, which remains the basis for all microelectronic technology.

Left: Jack Kilby's original notebook shows his notes and the two original integrated circuits; he was awarded the Nobel Prize for his work.

Kilby's invention something that would revolutionize the electronics industry.

Jack St. Clair Kilby was born in Jefferson City, Missouri and grew up in Great Bend, Kansas. After attending the University of Illinois at Urbana and armed with an MSc from the University of Wisconsin, he joined TI in mid-1958. As a new employee, he did not qualify for any vacation time. Left alone that summer, he re-examined the costs involved in producing

etched into it. This was known as the Integrated Circuit (IC) for which Fairchild Semiconductor filed a patent in July 1959, four months after Kilby.

Kilby and Noyce both share the credit for the invention of the integrated circuit, or microchip, and their companies patented different aspects of the design and manufacture. Yet, Noyce further revolutionized the semiconductor industry by co-founding Intel in 1968, which became the world's largest semiconductor company. Overseeing the invention and manufacture of the microprocessor, which was introduced in 1971, Noyce also helped define the Silicon Valley working style, eschewing the trappings of high corporate office.

Left: Founder of Fairchild Semiconductor, Robert Noyce further revolutionized the semiconductor industry by co-founding Intel in 1968, the world's largest semiconductor manufacturer.

Below: Intel continues to dominate the world of computing technology, launching ever-more powerful processing chips. The 45nm process 300mm silicon wafer was launched in 2007.

integrated amplifier circuits – a problem commonly called the "tyranny of numbers." Working with borrowed and improvised equipment, he conceived and built the first electronic circuit in which all of the components were built into a single piece of semiconductor material half the size of a paper clip. This solved the problem and was the conclusion he presented to the September gathering. A patent for the solid circuit was filed the following February.

In also inventing the handheld calculator and thermal printer, Jack Kilby was undoubtedly one of those with an insight matched by a professional talent, and by the time of his retirement in 1983, he held more than 50 patents. He was awarded the Nobel Prize for Physics in 2000.

Robert Norton Noyce was a confident physics graduate at Grinnell College, who went on to gain his doctorate at Massachusetts Institute of Technology (MIT) in 1948. He then worked at Shockley Semiconductor, working with William Shockley, the man credited with the invention of the solid-state transistor. Shockley and Noyce's scientific vision and egos soon clashed, and in 1957 Noyce left to co-found founded Fairchild Semiconductor. Like Kilby at TI he, too, came up with the idea of a single wafer or chip of silicon with many transistors all

Martin Luther King

1929–1968

American

Civil rights campaigner and Baptist minister

Martin Luther King was the major figure in fighting for African-American civil rights in the 1950s and 60s, which won him the Nobel Peace Prize in 1964, and ultimately ended the legal segregation of black and white people in the USA.

King was born in Atlanta, Georgia, into a middle-class family in the black community, and following in his father's footsteps, he chose to take up a career in the church. During the first half of the 1950s, he married and became pastor at the Dexter Avenue Baptist Church in Montgomery, Alabama.

Montgomery housed a small group of civil rights activists and in 1955 they decided to contest the city's racial segregation laws on the transport system. Rosa Parks refused to give up her seat on a bus to a white man and was arrested for her actions. Young, competent and eloquent, King organized a black boycott of the bus system that lasted for over a year and became a bitter dispute. But the following year elicited a ruling from the Supreme Court declaring the segregation on transport to be unconstitutional.

Spurred on by the success of the bus boycott, and recognizing the need for a wider platform for civil rights' campaigning, King founded the Southern Christian Leadership Conference in 1957, espousing a non-violence policy in a manner similar to Gandhi in India and, initially, the ANC in South Africa.

From 1960–1965, King's leadership and influence were at their most powerful. He organized marches, protests and direct action against obvious areas where African-Americans were denied access. Arrested many times, King was himself the target of violent attack. In 1963, Birmingham, Alabama became the centre for a series of boycotts and sit-ins, leading to a loosening of the segregation laws in that city. The March on Washington in August 1963 united the various civil rights organizations of America, when more than a quarter of a million people protested in the nation's capital. With his famous speech beginning, "I have a dream," King outlined his vision for a future America with equality for all – oratory that ranks with Lincoln's Gettysburg Address.

The result was the Civil Rights Act of 1964, which authorized the federal government to outlaw discrimination and to authorize desegregation. In December that year, King was awarded the Nobel Prize for Peace.

In 1965, the Selma to Montgomery marches provoked a violent reaction from the police and King's policy of non-confrontation lost him the support of more radical followers. Impatient with the slow pace of change despite the Civil Rights Act, in 1968, King organized the "Poor People's Campaign" which highlighted the desperate economic state of many black families.

In March 1968 King traveled to Memphis, Tennessee to support striking black sanitation workers. He had arrived safely in the city despite a bomb threat to his plane, but the evening of April 4 he was killed by an assailant later identified as James Earl Ray. King's voice may have been silenced, but his legacy still looms large in America. A charismatic man and effective campaigner, his work in the cause of civil rights hanged America forever.

Above: Eloquent and charismatic, King galvanized the civil rights movement and through a policy of non-violence achieved real gains for underprivileged Americans.

Below: A get-well letter from Richard Nixon, from the collection of Dr. Martin Luther King Jr.

Above: Martin Luther King waving to supporters on the Mall, Washington DC, during the March on Washington, August 28, 1963.

Below: In December 1964, at the age of only 35, King received the Nobel Peace Prize in Oslo, Norway.

OFFICE OF THE VICE PRESIDENT

WASHINGTON

September 22, 1958

Dear Dr. King:

I was terribly distressed to learn of the attack that was made on you in New York Saturday. To have this incident added to all of the unfortunate indignities which have been heaped upon you, is indeed difficult to understand.

I can only say that the Christian spirit of tolerance which you invariably display in the face of your opponents and detractors will in the end contribute immeasurably in winning the support of the great majority of Americans for the cause of equality and human dignity to which we are dedicated.

Mrs. Nixon joins me in sending our best wishes to you and Mrs. King.

Sincerely,

Richard Nixon

Dr. Martin Luther King, Jr.
Harlem Hospital
Lenox Avenue and 136th Street
New York, New York

Paul Lauterbur, Sir Peter Mansfield

Paul Lauterbur

1929–2007

American

Chemist

Sir Peter Mansfield

Born 1933

British

Physicist

Paul Lauterbur and Peter Mansfield shared the 2003 Nobel Prize for Physiology or Medicine for their work that made possible the development of magnetic resonance imaging (MRI) as an invaluable, non-invasive, diagnostic method.

An MRI scan is a safe, painless method of providing detailed three-dimensional pictures of organs and other soft tissue structures in the human body. The scanner uses a tunnel-like instrument about six feet long that is surrounded by a large circular magnet. The patient is moved into the hollow scanner and a strong magnetic field is used along with radio waves to align protons within the hydrogen atoms in human tissue. The realigned protons then transmit radio signals to a receiving device linked to a computer. Based on these signals, the computer then creates clear pictures of most parts of the body that can be used where other tests (such as X-rays) do not give medical experts enough information.

Paul Christian Lauterbur was born and raised in Sidney, Ohio. After high school chemistry experiments of his own, he was drafted into the US Army in the 1950s, where his superiors also allowed him to spend time working on a nuclear magnetic resonance (NMR) machine. He went on to graduate in 1962 at the University of Pittsburgh and credits the idea of the MRI to a brainstorm one day at a restaurant. He scribbled down an idea of the MRI's first model on a table napkin and used in further research at New York's Stony Brook University.

However, attempts at publishing his findings met with rejections. Several years passed before his pioneering work throughout the 1970s eventually came to be recognized and combined with Peter Mansfield's developments in the UK at the University of Nottingham.

Born in1933 in the south London working class area of Lambeth, Peter Mansfield left school unqualified at 15. After early work as a printer's assistant and in the UK Rocket Propulsion Department, post-National Service studies gained him a place at Queen Mary College, University of London. Studying physics, he chose as his undergraduate projects the building of a portable NMR spectrometer using new transistorized technology. Post doctorate work in the USA at the University of Illinois at Urbana in 1962 involved NMR study of doped metals. He returned to the UK, working for

Left: Paul Lauterbur shared the Nobel prize for medicine with Peter Mansfield for work which led to the development of MRI machines.

同一个世界 同一个梦想
One World One Dream

Above: The Magnetic Resonance Imaging scanner made detailed medical investigations possible without using invasive surgical technique or damaging X-rays.

Right: Sir Peter Mansfield at the Sir Peter Mansfield Magnetic Resources Centre, Nottingham University.

several decades at the University of Nottingham, along with valuable sabbaticals abroad, During his MRI researches, he became aware of Lauterbur's earlier work in the USA and took it a step further, combining it with his own researches to develop a mathematical process that speeded up image reading so that MRI could be used to produce clear images of the body.

Mansfield retired from teaching at Nottingham in 1994, but has continued research into MRI safety matters. Lauterbur died of kidney disease at his home in Urbana in March 2007. But their influence continues to be felt around the world every day, each time an MRI scan is used in the saving of yet another human life.

Vladimir Lenin

1870-1924

Russian

Russian revolutionary and founder of the Soviet Union

Lenin, born Vladimir Ilich Ulyanov, was one of the leading political figures and revolutionary thinkers of the 20th century. He masterminded the Bolshevik takeover of power in Russia in 1917, establishing the first Communist government in the world. He took on the name Lenin after a period of exile in Siberia at the turn of the 20th century.

Born into a well-educated family in Simbirsk on April 22, 1870, Vladimir was an excellent if rather argumentative

Below: Lenin (center) walking across Red Square, Moscow, surrounded by his Soviet military commanders, on May 25, 1919.

scholar; he did well in his exams and went on to study law. Lenin was deeply influenced by the revolutionary political views of his older brother, Alexander, who introduced him to the works of Karl Marx. Lenin went to Kazan University where he was exposed to further radical thinking. The arrest and execution of Alexander for plotting to kill the Tsar of Russia undoubtedly had a profound effect on Lenin, and in the same year he was expelled from university for participating in a student demonstration.

In 1890 Lenin moved to St Petersburg and completed his law degree and became a professional revolutionary. On his return from a trip to Europe he was arrested and imprisoned for carrying Communist books and leaflets, which were strictly forbidden in Russia. After 15 months he was sentenced to three years exile in Siberia. While living in exile Lenin wrote *The Development of Capitalism in Russia* and *The Tasks of Russian Social Democrats*.

On his release, Lenin spent much of the following 15 years

in Western Europe becoming the main leader of the Social Democratic Party. He produced a newspaper called *Iskra* ("The Spark") that was smuggled into Russia by his supporters. However, Lenin's uncompromising views – he insisted that membership be limited to a small core – caused a split in the party; his supporters were thereafter referred to as Bolsheviks and his opponents known as Mensheviks.

The 1905 St Petersburg Massacre, when Cossacks fired on peaceful protestors, spurred Lenin to advocate violence, all of which resulted in a number of uprisings. Lenin himself returned to Russia for two years but his hoped-for revolution was quelled when the Tsar Nicholas II made sufficient concessions to appease the Russian people. In 1907 Lenin went into exile again and remained in Western Europe until 1917. In 1916 he published *Imperialism, The Highest Stage Of Capitalism*, attributing war to capitalist nations' need to constantly expand.

By 1917, Russia was exhausted by the First World War

Above: After Lithuania declared independence from the Soviet Union in 1990 the statue of Lenin in Vilnius was dismantled.

Left: A lifelong revolutionary and Communist, Lenin, led the October 1917 Communist revolution in Russia.

and was ripe for change. Lenin returned home to work against the provisional government that had overthrown the Tsarist regime. In October of that year, having convinced a majority of the Bolsheviks to seize power, Lenin founded the Socialist State. Three years of civil war followed, during which time Lenin chillingly crushed any opposition, before the Bolsheviks assumed total control of the country.

Lenin survived an assassination attempt in 1918, but his long-term health was affected. He suffered a stroke in 1922, the same year that the Union of Soviet Socialist Republics (USSR) was formed, but died in January 1924. St Petersburg was known as Leningrad from 1924 until 1991 when it reverted to its former identity after the collapse of the Communist regime.

Abraham Lincoln

1809–1865

American

16th President of the USA

Abraham Lincoln presided over his country during its most turbulent period since Independence. He successfully preserved the Union of states despite the bitter Civil War, and successfully pushed through the legislation to abolish slavery. These actions, and his eloquence in defense of democracy have ensured that he is regarded as one of the great American presidents.

Lincoln was born in a log cabin in Hardin County, Kentucky in 1809, and by 1816 the family had moved to Indiana living a life Lincoln described as being far from easy. His father was virtually illiterate and, by his own admission, the young Abe's own schooling could be best regarded as minimal, although he did learn to read and write.

In 1830, the family moved to Illinois where a rift opened up between father and son, leading to Abe's departure to New Salem near Springfield. The younger Lincoln worked at a number of occupations in a short space of time and he served as an officer in a volunteer company during the Black Hawk War of 1832. In 1834 he was elected to the Illinois legislature, and two years later, in 1836, having taught himself grammar and mathematics, he passed the exams enabling him to practice law.

Above: Lincoln was president throughout the Civil War and had been elected partly because of his opposition to extend slavery.

Left: The contents of President Abraham Lincoln's pockets on the night he was assassinated by John Wilkes Booth.

He set up a partnership with a fellow Whig in Springfield in 1837 but other issues were starting to dominate his life. He married in 1842 and took his first step into national politics in 1847 when he started a two-year term in the US House of Representatives. Thereafter, politics took a back seat in Lincoln's life, for he ploughed his energies into his law practice. By 1856 however, he had joined the fledgling Republican Party but failed in a bid to make the US Senate two years later.

Undaunted, he continued on his political path and his debates with Stephen Douglas over the pro-slavery Kansas-Nebraska Act won him national fame. He was nominated as the Republican candidate for the 1860 presidential election, which he won.

The Southern states resented federal "'interference" in their way of life and particularly in their economy, much of which revolved around slave labor. Before Lincoln had even been inaugurated, South Carolina seceded from the Union, the first of six states to form the Confederacy. A month later, in April 1861, the first shots of the Civil War were fired at Fort Sumter.

Lincoln was a strong commander-in-chief, fully participating in military planning until he handed over to the eminently capable General Ulysses S. Grant in 1864. The Emancipation Proclamation of 1862 freed slaves in the Confederate states and compensated their owners, signaling that the abolition of slavery was a northern war aim. In November 1863, at the dedication of the Soldiers' National Cemetery of Gettysburg, Lincoln made clear another vision of "government of the people, by the people, for the people," justifying the sacrifices of the war in the Gettysburg address.

Re-elected in 1864, Lincoln was shot on April 14 1865, only five days after Robert E. Lee surrendered to Ulysses S. Grant at Appomattox. He died the following morning.

Above: An idealized portrait of Lincoln writing the Emancipation Proclamation, which declared that all slaves in rebel held territory would be "thenceforward, and forever, free."

Below: Lincoln (easily distinguishable from his height and bearded profile) visiting soldiers encamped at the battlefield of Antietam in Maryland, in October 1862. Lincoln did not enter the war lightly, but proved a capable commander-in-chief.

Charles Lindbergh

1902–1974

American

Aviator

Charles Lindbergh became an instant hero in May 1927 after he made the first solo non-stop flight across the Atlantic in his monoplane *The Spirit of St Louis*. He was a renowned inventor, explorer and author, as well as a record-breaking pilot, and remained an advisor to the aviation industry from his early flying days through to the age of supersonic jets.

Born in 1902, Lindbergh was raised on a Minnesota farm. He enrolled at the University of Wisconsin to study engineering in 1920, but within two years he dropped out to become a daredevil pilot or barnstormer, performing stunts at county fairs. When, aged 22, he enlisted in the US Army Air Service, Lindbergh was already a skilled pilot. By his mid 20s he had racked up hundreds of hours of flying time and had gained a reputation locally as a decent pilot.

In 1919 Raymond Orteig, a New York hotel owner, had offered $25,000 to the first aviator who could fly non-stop from New York to Paris. Eight years on, the prize was still unclaimed and Lindbergh, far from being put off by knowing several others had died in the attempt, was convinced he could do it. Having persuaded nine St Louis businessmen to share the expense of building a purpose-built plane, which he helped design, the record-breaking attempt began to take shape.

The plane, manufactured by the Ryan Aeronautical Company, was a M-2 strut-based monoplane; Lindbergh named it *The Spirit of St Louis*. On May 20, 1927, the unknown Lindbergh set off from Roosevelt Field, near New York City; some 36,000 miles and 33.5 hours later, 'Lucky Lindy' as he would be dubbed, landed at Le Bourget, near Paris, an international star.

For the next five years Lindbergh received numerous awards and celebrations, with President Calvin Coolidge giving him the Congressional Medal of Hour and the Distinguished Flying Cross. Lindbergh flew many goodwill flights at the request of the US government, including a tour of Latin America. While in Mexico he met Anne Spencer Morrow, daughter of the US ambassador; the pair married in 1929.

Sensational headlines shocked the world in 1932 when the couple's 20-month-old son, Charles Jnr., was kidnapped and murdered. To escape the lurid publicity, Lindbergh and his wife moved to England. Around that time, Lindbergh collaborated with Alexis Carrel, a French surgeon and biologist, in inventing an "artificial heart and lungs," a perfusion pump to keep organs alive outside the body by supplying blood and air to them.

Above: Lindbergh's 1954 autobiography, **The Spirit of Louis**, is a literary account of his epic flight, which won him a Pulitzer Prize.

Lindbergh's iconic status was tainted when, after conducting various air-power surveys in Europe in the late 1930s and having been greatly impressed by the capabilities and scope of the German aircraft industry, he warned that United States involvement in World War II could not prevent a German victory. In 1938, Hermann Goering had even presented him with a German medal of honor and some Americans accused Lindbergh of being a Nazi sympathizer when he refused to return the medal.

After the war his reputation was restored with the publication in 1954 of *The Spirit of Louis*, an expanded account of his epic flight, which won a Pulitzer Prize.

Above: Lindbergh with his first wife, writer Anne Morrow Lindbergh, shortly after their marriage in May 1929.

*Below: Charles Lindbergh made the first solo non-stop flight across the Atlantic in May 1927 in the biplane the **Spirit of St Louis**.*

Joseph Lister

1827–1912

British

Surgeon and medical pioneer

Above: Lister as a young man. He was knighted by Queen Victoria in 1883 for pioneering antiseptic surgery and hygienic medical practice.

Joseph Lister transformed the medical world by successfully promoting the idea of sterile surgery in the 1860s. It was a major medical breakthrough: cleaner, antiseptic operating conditions led to a vast improvement patient survival from major surgery.

Born into a Quaker family in 1827, Lister was the fourth child and second son of British Physicist Joseph Jackson Lister, inventor of the achromatic lens for microscopes, a device that elevated them to the status of serious scientific tools. The family was based in Upton, Essex, and Lister junior attended Quaker schools in Hertfordshire and London at which science subjects were high on the agenda. He also became fluent in French and German, the leading languages of medical research.

Lister was an excellent student, but his non-conformist religious beliefs meant he was barred from studying at Oxford and Cambridge universities. Lister graduated as a bachelor of medicine with honors in 1852. A year later, a move to Edinburgh not only had a profound effect on Lister's life but also on the medical world. He traveled to Scotland to spend a month with Professor James Syme, regarded as the greatest teacher of surgery at the time. Lister decided to stay on as Syme's assistant (he would remain in Edinburgh for 24 years) and within three years had married his mentor's daughter, Agnes. Theirs was a long and happy marriage, and Agnes gave great professional support to her husband, helping with laboratory experiments and writing up his notes.

When he became a surgeon, Lister was horrified that more patients died after operations than during them. In the Edinburgh hospital where he worked, almost 50% of all surgery patients died from infections, and in some parts of Europe, the figure was nearer 80%.

Influenced by Louis Pasteur's research that decay was caused by living microorganisms in the air, Lister considered that microbes in the air were probably causing the wounds to putrefy and had to be destroyed before they had the chance to do so. Lister's methods of cleaning and dressing wounds with a solution of carbolic acid initially met with indifference and even hostility, but the occurrence of gangrene declined markedly. He insisted that surgeons wore clean gloves and also introduced the sterilization of surgical instruments using heat and carbolic acid, as well as the frequent cleaning of the surgeons' gloved hands with carbolic acid solutions during operations.

In 1867 Lister announced to a British Medical Association

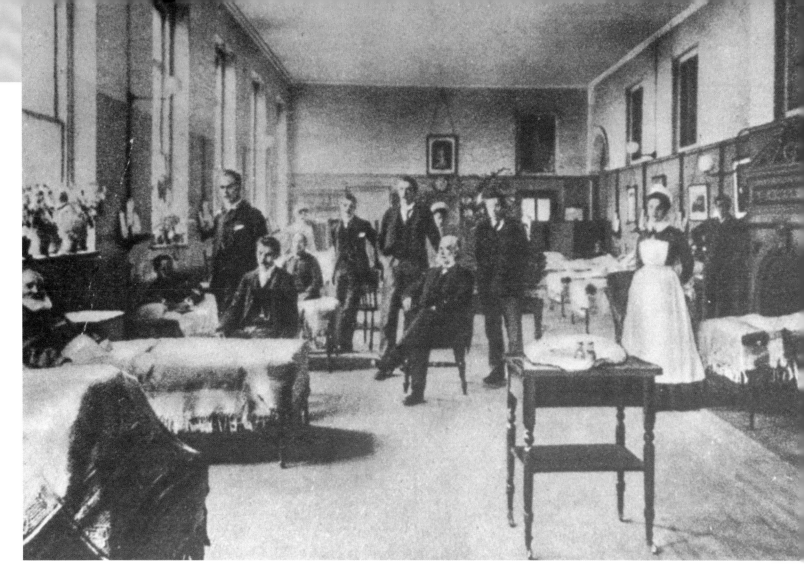

Above: Lister (seated) at King's College Hospital, where he held the chair of surgery. His insistence on sterilizing instruments and treating wounds with antiseptic led to a dramatic decline in deaths from infections.

Right: Lister and Louis Pasteur receiving honors from the Sorbonne, Paris, in 1885. These two great scientists transformed medicine in the 19th century and were great friends as well as medical pioneers.

meeting that his wards at the Glasgow Royal Infirmary had been sepsis-free for nine months. Gradually his theory of became more widely accepted, although doctors from Denmark and Germany implemented the antiseptic principle before those on Lister's home soil. It was not until he was appointed Professor of Surgery at London's King's College Hospital in 1877 that English doctors fully appreciated Lister's work. Within two years antiseptic surgery was universally acclaimed.

Queen Victoria knighted Lister in 1883; in 1897 he became Lord Lister of Lyme Regis, the first Briton to be honored for services to medicine.

Martin Luther

1483–1547

German

Theologian and religious reformer

Left: Luther controversially denied the authority of the pope and stressed that man's salvation comes from faith alone.

Martin Luther was the humble German cleric who challenged Christianity by daring to question the authority of the Pope and the Roman Catholic Church. In doing so he became one of the most influential figures in the history of Christianity, a founder of Protestantism and a leader of the Reformation.

Luther studied at an Augustinian seminary, where he lived the life of a monk and became an ordained priest in 1507. But he became disillusioned with the ecclesiastical abuses he saw all around him. He suffered a crisis of faith, but finally came to believe that salvation in heaven could only be achieved by faith in God and not by actions or even good works; a person could only be saved through God's infinite love and mercy alone, which could not be bought with material goods. In an age when the Catholic Church sanctioned the purchase of pardons and the sale of indulgences as a means to raise money, Luther's austere attitude was bound to be unpopular.

He itemized the corruptions of the church, attacking doctrines and practices that he believed to be wrong by nailing his "Ninety-five Theses" to the door of Wittenberg church on October 31, 1517. Intended to be a submissive, questioning document rather than accusatory and intolerant, it was distributed, initially locally, and then around Europe. At this stage, Luther was not trying to split the Church, merely to drive out corruption.

Seeing itself as under attack, the Catholic Church hierarchy, led by the venal Pope Leo X, fought back by trying to discredit Luther, but, similarly sickened by the overweening corruption of the Church many people supported Luther's stance. Luther decided that the Bible was the true source of doctrinal authority and renounced obedience to Rome. By 1519 Luther was labeled a heretic and Leo X issued a papal bull, *Exsurge Domine* ('Lord, Cast Out') condemning Luther and his reforming ideas. A few months later Luther was excommunicated. Luther responded by burning his copy of the papal bull.

On April 17, 1521 Luther answered the summons to appear before the Diet of Worms to explain himself. Ordered to recant, he refused, saying, "Here I stand. I can do no other. God help me. Amen." The Edict of Worms declared Luther and his followers to be outlaws, but he was offered sanctuary by

Frederick III, Elector of Saxony. With his freedom somewhat curtailed, Luther set about translating the Bible, the true source of Christian truth in his mind, into German, so his fellow countrymen could read the words for themselves and make up their own minds. Additionally, this work helped to standardize German as a single language. The precedent of making the

Right: A statue of Martin Luther outside the Church of Our Lady in Dresden, Germany.

Below: Martin Luther nailing his incendiary "Ninety-five Theses" to the door of Wittenberg church on October 31, 1517.

Bible available in the vernacular was highly significant to the emerging Protestant cause.

Luther's work was continued after 1522 by more radical clerics, such as Zwingli, and the cause of reform became mixed up in politics. Luther continued to play a part, writing pamphlets and books explaining his position and promoting the Protestant cause. When he died in 1547 the Protestant Reformation was already changing the world and other, more extreme, Protestant thinkers were coming to prominence.

Madonna

Born 1958
American
Musician and actor

Madonna has been called "one of the greatest pop acts of all time," and ranked as the best-selling female rock artist of the 20th century, yet, she has always courted controversy, enjoying both adulation and disparagement in equal measure. Her popularity – and astute business sense – has enabled her to exert a level of power and control unusual for a woman in the entertainment industry.

Born on August 16 1958 in Bay City Michigan, Madonna Louise Vernon Ciccone was one of six children of a French Canadian mother and first-generation Italian father. She moved to New York in 1977 to pursue a career in dance. However, after a singing with pop bands, she released her debut album and followed it with four, consecutive number-one albums.

Right: Comparatively early in her career, Madonna performed to a sold-out crowd at JFK Stadium, Philadelphia for the Live Aid concert on July 13, 1985.

Below: In costume as Eva Perón for the movie adaptation of the musical Evita *with the director Alan Parker.*

Her music incorporated catchy tunes and breezy lyrics rooted in love and relationships, often with earthy eroticism added for an extra frisson. "Like a Virgin" released in 1984 provided her first US number one, and was swiftly followed by "Material Girl" and another nine chart-topping singles. She was the first female artist to fully realize the appeal of music videos, which she used to give her songs an additional visual punch. Madonna is known for exploring religious symbolism and sexual themes in her work, and certainly enjoys challenging her audiences' preconceptions. Madonna's oeuvre has been attacked with as much zealous intensity as it has been praised by her fans, and she has becoming the subject of academic interdisciplinary research. Globally, publications examine her place within popular culture, analyze her mass-media spectacles, and study the iconography of sexual minorities in her videos.

After the public and commercial success of her "Blonde Ambition" tour in 1990, in 1992 the "Queen of Pop" formed her own entertainment company, Maverick, enabling her to exert full control over her work and image. She published a book of photographs, starkly entitled *Sex*, along with a promotional video of blatantly erotic themes. Predictably controversial, these generated negative publicity, and Madonna seemed to react by mellowing slightly, starring in *Evita* in 1996 and giving birth to her first child. Her recording career bounced back in 1998 with the release of her critically acclaimed album *Ray of Light*.

As well as her hugely successful recording career, Madonna has acted in 22 films, although most failed critically and commercially. But films were the medium through which she met, married and divorced actor Sean Penn. Her daughter Lourdes Maria ('Lola') in 1996 was fathered by her personal trainer Carlos Leon, before she married British film director Guy Ritchie. With him she acquired a country estate in southern England, their eight-year marriage producing two sons, Rocco and David Banda, the small boy they adopted from Malawi.

Madonna's ever-changing image and the impact of modern media means that no other female has had such a rapidly widespread influence on young women throughout the world. By 2008 Madonna had become the most successful female in the UK singles chart history; she surpassed Elvis Presley as the artist with most Billboard Top Ten hits; and she was inducted into the Rock and Roll Hall of Fame. In addition, her "Sticky & Sweet" stage act was the highest earning concert tour by a solo artist.

Above: Madonna on stage at the Morumbi stadium in Sao Paulo, Brazil during part of her "Sticky and Sweet" tour in 2008.

Malcolm X

1925–1965

American

African–American political activist and civil rights leader

Articulate, charismatic and radical, Malcolm X was a prominent figure in the US civil rights movement, who advocacy of Black Nationalism made him an iconic figure and provided the intellectual basis for the Black Power movements of the 1970s.

Malcolm X was born Malcolm Little in Omaha, Nebraska, and his father, a Baptist preacher, was active in African-American civil rights at a time of considerable racial turmoil. Malcolm had an impoverished childhood and experienced racial hatred from an early age. The Ku Klux Klan burnt down the family home in Lansing, Michigan; his mother was placed in a mental institution; his father was killed in unclear circumstances, and he and his siblings were split up. Malcolm settled in Boston with his half-sister but turned to crime. He was convicted of burglary in 1946 and sentenced to 10 years in jail.

It was in prison that Malcolm turned his life around. He studied the work of the National of Islam (NOI), a radical Muslim organization that argued for African-American separation

Above: Malcolm X insisted that African-Americans did not need integration, but rather powerful self-governing institutions able to withstand racial violence.

from the wider US community and demanded absolute discipline from its adherent. Under its guidance, he read prodigiously, including the Koran, the Bible and various books on history and literature. The former petty thief was released in 1952 and, now radicalized, rejected his family name in favor of Malcolm X. He quickly rose through the rank of the NOI, becoming the minister of Harlem's Temple No. 7 two years later. Malcolm often spoke out against the more mainstream civil rights movement, arguing that African-Americans did not need integration, but needed powerful self-governing institutions that could stand up to racial violence.

Malcolm also spoke out against some of the core tenets of the NOI, chiefly its unwillingness to take part in mainstream politics. He was willing to oppose the movement's leadership, which centered on Elijah Muhammad, and Malcolm participated in or supported various civil rights protests. Malcolm was also an outstanding orator and was a genuinely popular figure, who appeared to increasingly challenge the fundamental beliefs of the NOI's leadership. A rift was inevitable, and in 1964, shortly after Elijah had been revealed to be the father of two children by two of his ex-secretaries, Malcolm announced his resignation. He went on to form the Muslim Mosque, Inc., which was dedicated to the wider political process and the mainstream civil rights movement.

That same year he

Above: In 1964 Malcolm changed his name to El Hajj Malik El-Shabazz and announced the foundation of the Organization of Afro-American Unity.

Right: Malcolm X addressing a meeting in 1964, the year he resigned from the Nation of Islam and formed Muslim Mosque, Inc.

visited Mecca and paid a second trip to several Arab countries and others in Africa. On his return to his homeland, Malcolm announced that was henceforth to be known as El Hajj Malik El-Shabazz and that he was founding the Organization of Afro-American Unity (OAAU), a body that argued for independent African-American institutions, such as schools, but also backed participation in the wider political process. The OAAU was short-lived as Malcolm was assassinated by gunmen associated with the NOI in New York City on February 21, 1965 and his new movement quickly disappeared without his charismatic leadership. Despite this sudden end to the OAAU, there is little doubt that Malcolm X was the most significant leader of the African-American nationalist movement in the postwar United States.

Nelson Mandela

1918-2013

South African

South African statesman and president

Above: Mandela traveled outside South Africa to gain support for the ANC. Here he speaks at a conference in c.1960, before he was convicted of sabotage in October 1962.

Left: Mandela maintained a high public profile into old age. Here he acknowledges the Melbourne crowd after speaking at the Colonial Stadium for the World Reconciliation Day

Nelson Mandela emerged over a number of decades as South Africa's major opponent to apartheid, survived the deprivations of prison life over quarter of a century, and subsequently became the first black president of his country. Mandela's support for reconciliation within South Africa after many years of racial dislocation has helped to heal the divisions and forged a transition towards multiracial democracy.

Born in the Transkei in 1918, Mandela renounced his hereditary rights as the son of a chief to study law. He attended the University College of Fort Hare from which he was suspended for taking part in a protest boycott. Undaunted by this experience he went to Johannesburg, completed a Bachelor of Arts degree and entered into his studies for the legal profession. In 1942 he became a member of the African National Congress (ANC), the political party that would fight

to end the apartheid regime in South Africa.

After the 1948 South African election, which was won by the all-white, apartheid-supporting National Party, the ANC constituted a Program of Action, partly drawn up by Mandela, which urged a peaceful policy of opposition to the new government in the shape of strikes, boycotts and civil disobedience. As a result of his support for the Defiance Campaign of 1952, Mandela received a suspended prison sentence, a ban on attending any kind of gathering and a six-month order confining him to Johannesburg.

After the Sharpeville Massacre of 1960, the ANC was banned. Renouncing his commitment to peaceful protest, in 1961 Mandela became leader of the ANC's armed wing, Umkhonto we Sizwe, a decision he described in his autobiography as a last resort. In 1962 Mandela received a jail sentence of five years, after his return from a military training trip abroad. That five-year stretch became a life sentence after he was found guilty of sabotage during the contentious Rivonia trial in 1964.

Until his release in 1990, Mandela became the high profile inmate of Robben Island prison, the driving force behind

the education courses he urged upon his fellow prisoners and, despite his incarceration, the symbol of defiance against apartheid. On more than one occasion he refused the chance of release, as it would have compromised his principles.

With the apartheid regime widely condemned around the world, Mandela was finally released in 1990. In 1991 he became president of the ANC and worked with the South African president F. W. de Klerk to effect a peaceful transition to nonracial democracy within South Africa. In 1994 Mandela was elected president in the first democratic election, winning 66% of the vote. A strong advocate of reconciliation, he headed a government of national unity, and established the Truth and Reconciliation Commission to investigate human rights abuses during the apartheid years.

When he retired from political life in 1999 he still remained the most respected figure in South African life and it was perhaps no surprise that, despite his age, he still made an appearance for the ANC during the 2009 elections. His benevolence, honesty, and charisma have made him an almost iconic figure throughout the world, and an international figure of great moral stature.

Above: February 11, 1990 Mandela was released from Victor Verster prison (where he spent the last three years of his captivity). He was met by his wife, Winnie, and exulting crowds.

Below: Mandela with players during the Springboks and Bafana Bafana visit to the Nelson Mandela Foundation in Houghton, Johannesburg, South Africa.

Guglielmo Marconi

1874–1937

Italian

Inventor/entrepreneur

Guglielmo Marconi was an Italian inventor and electrical engineer whose groundbreaking invention was the wireless telegraph, the forerunner of present-day radio. He was awarded the Nobel Prize for Physics in 1909.

Marconi was born in Bologna on April 24, 1874. His father, Guiseppe, was an Italian landowner, while his mother, Annie Jameson, hailed from Ireland. Marconi was educated at the Technical Institute of Livorno and went to the University of Bologna. He became fascinated with the work of German physicist Heinrich Hertz, who had been experimenting with electromagnetic waves.

Working on his father's estate in 1895, Marconi began experimenting himself. He improved on Hertz's ideas and, having discovered that an insulated aerial enabled him to increase transmission distance, succeeded in sending wireless signals not only over a hill, but over a distance of up to two miles.

Marconi patented his work but the Italian government saw no value in it, so he went to England where he gained support from the British Post Office and founded his own wireless telegraph company. It was a huge gamble – Marconi borrowed more than £50,000 from British banks in order to make the world sit up and take notice of his invention. He managed to secure contracts with several shipping lines, installing transmitters in the navies of Britain, France, Germany and Italy, and then earned a contract to provide wireless telegraphy to the United States.

Marconi could see that the whole exercise was not only potentially highly profitable, but also of great benefit to society. And yet things did not go entirely his way, with high winds playing havoc with the towering antennae constructed in England and the masts across the Atlantic in Cape Cod. In 1899, Marconi had managed to send a signal across the nine-mile expanse of the English Channel to France. But now, if the blustery winds behaved themselves, he was set to achieve something far more dramatic.

Above: Marconi turned his wireless invention into a successful global business as well as also patenting a number of other inventions.

Most people considered the Earth's curvature would prevent the passage of signals over vast distances. Marconi, though, forged ahead and in December 1901, despite further breezy conditions, he managed to send the first ever transatlantic signal from Poldhu in Cornwall, England to St. John's in Newfoundland – a distance of 2,100 miles.

Marconi was able to develop and finance an expansion of his businesses given his family's good connections and, during the early years of the 20th century, patented a number of inventions, including what became the standard wireless receiver for many years. In 1909, the Italian won the Nobel

Above: In 1909 Marconi shared the Nobel Physics Prize with Karl Braun, who modified his transmitters to increase their range and strength.

Right: Marconi (center) with his assistants Kemp (left) and Paget at Signal Hill, Newfoundland, shortly before receiving the historic transatlantic signal.

Prize in Physics, an award shared with Karl Ferdinand Braun, whose modifications to Marconi's transmitters dramatically increased their range and abilities.

During the First World War Marconi served in the Italian armed forces and directed his country's radio service. He began experiments with shortwave radio and narrow-beam transmission, making detection by the enemy much harder; once war was over, he was an Italian signatory for the 1919 Paris Peace Conference. In the years that followed Marconi continued his shortwave investigations and later became an active supporter and member of Mussolini's fascist government. The supreme inventor was given a state funeral after his death, from a heart attack, in 1937, when there was also a two-minute silence by all radio stations around the world.

Karl Marx

1818–1883

German

Philosopher and political theorist

Karl Marx was the major figure to emerge from the socialist movements of the 19th century and his ideas became the inspiration behind the majority of Communist states that evolved during the following century. His writings form the basis of the political movement known as Marxism.

Strange as it may seem, Marx, despite his views, was born in Trier, into a middle-class Jewish family, his father being a lawyer who had himself baptized as a Protestant to protect both his career and his standing in the community. At the age of 17 Marx seemed destined to follow his father into the legal profession, as he was accepted at the Faculty of Law at the University of Bonn.

During a four-year stint at the University of Berlin, he became engaged to Jenny von Westphalen but, more importantly from an historical point of view, took up with the Young Hegelian movement who followed the teachings of philosopher Georg Hegel. The thrust of ideas emerging

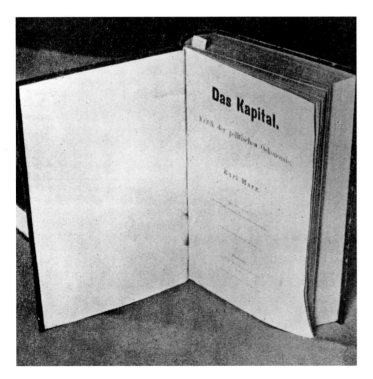

Above: A German-born Jew, Marx lived an itinerant life until he settled in London where he remained for 34 years.

Left: The title-page of volume one of Das Kapital. *Published in 1867, the work became the inspiration for modern international communism.*

from this clique included an opposition to Christianity and a condemnation of their Prussian rulers.

In 1842 Marx worked in Cologne editing the liberal newspaper *Rheinische Zeitung,* a short lived appointment, as his regular articles highlighting economic problems were frowned upon by the Prussian government, which shut down the publication.

Shortly after his marriage in 1843, Marx moved was to Paris, then the center for socialist thought. During this short stay he declared himself to be a Communist, and called for an "uprising of the proletariat". Expelled from France, he traveled to Brussels with his friend Friedrich Engels, where he remained for three years, before his itinerant life style took him back to Paris and eventually to London in 1849. Marx and Engels (the son of an industrialist) collaborated on a number of socialist pamphlets and were commissioned to write a program for the secret Communist League. This became *The Communist Manifesto*, which became the bible for the socialist movement. Writing that all history is the history of class struggle, Marx and Engels predicted the demise of capitalism and the eventual victory of the working class.

Marx scraped an existence in London, often supported by Engels, and while his writings became larger in ambition, his family's comfort was sacrificed to his ideological loyalties. By 1867 he was ready to publish the first volume of his most famous work, *Das Kapital*, which outlined at some length his philosophy on the exploitation of labor and his anticipated collapse of the capitalist system. Two further volumes were published by Engels after Marx's death.

In 1864 he had been elected to the general council of the First International, an organization of like-minded socialists dedicated to the spread of Communist philosophy. Although relatively short-lived, the First International set a precedent for the idea of worldwide co-operation between committed believers.

In later years Marx's health deteriorated to a great degree, although he continued with his writing, the content of which emphasized his unwavering belief that the downtrodden workers would inevitably triumph over their capitalist masters. Marx died on March 14, 1883 and is buried at Highgate Cemetery in North London. Speaking at the funeral, Engels declared, "At a quarter to three in the afternoon, the greatest living thinker ceased to think."

Above: Karl Marx's tombstone with his distinctive bust is one of the most visited memorials in Highgate Cemetery, north London.

Malcom McLean

1913–2001

American

Trucking and shipping entrepreneur

Malcom McLean changed the world shipping industry changed forever. On April 26, 1956 McLean loaded a ship with 58 35-foot containers and sailed from Newark, New Jersey, to Houston, Texas. The metal shipping container, which McLean developed, went on to replace almost completely the traditional "break bulk" method of handling dry goods, and revolutionized the transport of goods and cargo on a global scale.

He was born Malcolm Purcell McLean in Maxton, North Carolina on November 14, 1913. Later in life, he changed his given name to its historic traditional Scottish spelling of "Malcom." With only a high school education, McLean began work at a service station near his hometown, saving enough money by 1934 to buy a second-hand truck for $120. Together with his sister Clara and brother Jim, he founded the McLean Trucking Co. in nearby Winston-Salem, and started by transporting empty tobacco barrels and working as one of the drivers.

From that single pickup truck, he built the company into the second-largest trucking company in the United States, with 1,770 trucks and 32 terminals. Later, he said that the container-shipping concept came to him early when he had to cool his heels at Hoboken, New Jersey, waiting his turn to load bales of cotton on a ship. He realized it would save time and money if he could simply load his trailer onto a ship, and so in 1953 decided to get into the shipping business. However, it soon became apparent to McLean that his "trailerships" would be inefficient because of the large waste in potential cargo space on board the vessel – so

Left: Clara, Jim, and Malcom McLean (center) of McLean Trucking, photographed in Winston-Salem, North Carolina, in 1949.

he modified his original concept into loading on board just the containers, not the chassis – and containerization was born.

At the time, federal regulations would not allow a trucking company to own a shipping line, so McLean had to sell his trucking company before acquiring the Pan-Atlantic Steamship Corporation in the mid-1950s, which he used to launch the container venture in 1956, carrying containers on the East Coast. Pan-Atlantic became Sea-Land Service in 1960 and its international services were sold to Maersk in 1999, becoming Maersk Sealand.

McLean's concept of the "lift-on, lift-off" process – in which a gantry crane hoists containers – replaced the use of a wheeled chassis to transport loose cargo on and off vessels. This produced dramatic reductions in labor and dock servicing time. Determined to achieve industry standardization, and convinced that it was also the path to overall industry growth, McLean's efforts resulted in standard container designs.

Yet, although he was so influential in the growth of the world's economy, McLean died in relative obscurity at his home on the East Side of Manhattan in 2001, age 87. Fortunately, his contribution was recognized by the US Secretary of Transportation Norman Y. Mineta, who paid tribute, saying, "We owe so much to a man of vision, the father of containerization, Malcom P. McLean."

Above: McLean revolutionized the transportation of cargo with containerization and the "lift-on, lift-off" process, which reduced labor and dock servicing time.

Gregor Mendel

1822–1884

Czech

Scientist and geneticist

Above: Sadly Mendel's painstaking research and insights were not understood or valued in his lifetime, nor for several decades after his death.

Austrian monk and biologist Gregor Mendel worked quietly for years studying inherited characteristics in plants that became the basis of the modern theory of genetics. His work went virtually unnoticed for decades, until his description of the principles of hereditary transmission became known as "Mendel's Laws of Inheritance," and revolutionized not only the cultivation of plants, but also the breeding of domestic animals for desirable traits. Mendel started the study of genetics, but he also began the systematic use of mathematics, quantified measurements and applied statistics in biology. His work proved that evolution worked by predictable segregation ratios. Mendel's Laws specify the Law of Segregation and the Law of Independent Assortment.

Johann Mendel was born in 1822 in Heizendorf, Austrian Silesia (present day Hyncice, Czech Republic) into a penniless peasant farming family. The only way he could get a good education and escape a life of poverty was to enter the Augustinian Abbey of St Thomas at Brunn (Brno) in 1843, where he was given the name Gregor; happily for him it was a teaching order with a reputation for scientific enquiry. In 1851 he studied botany, chemistry, physics and zoology at the University of Vienna, but after failing to get a teaching diploma, he returned to the monastery in 1853, determined to conduct his own experiments with plants.

In 1856, with the necessary permission from his abbot, Mendel began investigating variation, heredity and evolution in plants using the quick growing common pea, *Pisum sativum* that already grew in the monastery garden. For eight years he grew and analyzed over 28,000 plants, comparing size, length, shape, growth and color. He protected them from accidental pollination and cross-pollinated them himself. He collected the seeds, grew them, bred them and compared them, again and

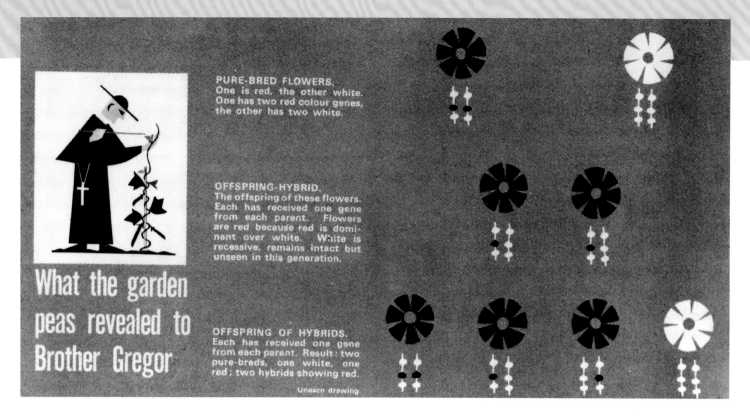

PURE-BRED FLOWERS.
One is red, the other white.
One has two red colour genes,
the other has two white.

OFFSPRING-HYBRID.
The offspring of these flowers.
Each has received one gene
from each parent. Flowers
are red because red is dominant over white. White is
recessive, remains intact but
unseen in this generation.

What the garden peas revealed to Brother Gregor

OFFSPRING OF HYBRIDS.
Each has received one gene
from each parent. Result: two
pure-breds, one white, one
red; two hybrids showing red.

Unesco drawing

Below: Sadly Mendel's painstaking research and insights were not understood or valued in his lifetime, nor for several decades after his death.

Above: An illustration showing Mendel's studies of garden pea hybrids that helped him develop the fundamental principles of modern genetics.

again all the time keeping copious notes and references. He began to see logic in the numbers and developed his concept of heredity units (now called genes) with either dominant or recessive characteristics. Through these methods he worked out the patterns and laws of heredity, for example that one in four plants will be purebred.

In 1865 Mendel read his paper — *Versuche über Pflanzen-Hybride* — to the Brno Natural History Society and then published his observations in their obscure journal, but his work was completely ignored. He even sent out 40 reprints to distinguished biologists. All but one, Carl Wilhelm von Nägeli, ignored him, and von Nägeli thought the work "incomplete." He was left deflated and discouraged, although still sure of the importance of his work. Two years later Mendel was elected abbot of St Thomas and became consumed by his heavy workload. Always intending to transfer his work to animals, he did a little breeding experimentation with bees, but their sting proved so lethal they had to be destroyed.

Mendel died age 61 in 1884 in Brno, an unrecognized genius and frustrated scientist: his successor burnt all his papers. Two decades passed until in the early 1900s his work started to be recognized for its significance in relation to evolutionary theory.

Dmitri Mendeleev

1834-1907

Russian

Chemist

Russian Dmitri Mendeleev played a vital role in the world of science by not only creating the first valid periodic table of elements, but also managing to leave suitable gaps in the table for elements yet to be discovered.

Born in February 1834 in Siberia, Dmitri Mendeleev was the youngest of at least 14 children; indeed, he may well have been the 17th child of Ivan and Marya Mendeleev. Dmitri's father, a headmaster of a local school, went blind not long after the birth of his youngest son and his mother, who came from a family that had introduced glass- and paper-making to Siberia,

had to return to work to support her family.

Dmitri showed great promise at school, displaying a fascination for mathematics, physics and geography. As he matured a brilliant intellect came to the fore, as did a sharp memory. He also learned a lot about glass and glass blowing from the family establishment. However, a series of domestic problems threatened to shatter this early promise.

First, his father died and then the family glass factory was destroyed by fire. A move to Moscow followed, from where the by-now needy family went to St Petersburg; Mendeleev, aged 16, was accepted at the Central Pedagogic Institute in 1850. However, further family hardship followed when his mother and a sister died from tuberculosis. Dmitri himself was later confined to bed with suspected tuberculosis but managed to maintain his studies thanks to his own fortitude plus consistent help from his teachers and fellow students who visited him to keep him on schedule.

And there was no doubting his brilliance. Aged 20, and having recovered from his illness, Mendeleev penned his first published scientific work, "Chemical Analysis of a Sample from Finland." By the time he was 22, Mendeleev had enjoyed his first teaching post, on the Crimean Peninsula, but with the Crimean War raging he returned to teach at the University of St Petersburg. Struggling to find an appropriate textbook for his chemistry teachings, he wrote his own. The two volumes of *The Principles of Chemistry* gave a perfect base for modern chemical and physical theory.

However, his real lasting legacy came with his first effort at formulating an effective periodic table of the elements in 1869. Mendeleev arranged the 63 known elements – the basic building blocks of matter – into a table based on atomic mass; his first table was compiled by arranging the elements in ascending order of atomic weight and grouping them by similarities in properties. He shrewdly foresaw the discovery of new elements, not only predicting their existence, but also the properties of those elements, leaving appropriate gaps for them in his table. By the 1880s, three elements had been uncovered, exactly as described, and Mendeleev's stock rose still further.

Mendeleev was widely honored by scientific groups all over Europe, yet he resigned from the St Petersburg University in 1890. He was almost 60 years of age and had been associated with the university for more than 40 years. In 1893 he was appointed Director of the Bureau of Weights and Measures and was soon making his mark – by the following year new standards of vodka production were introduced and in 1899 he introduced the metric system into Russia.

Mendeleev died in 1907 but his name lives on today. As well as his association with the periodic table of elements, a crater on the Moon is named after him, as is element number 101, the radioactive mendelevium.

Above: The Mendeleev Congress on General and Applied Chemistry has been held every four years since the first meeting in December 1907.

Left: Among many other achievements Mendeleev is credited with introducing the use of the metric system to the Russian Empire.

Opposite: Mendeleev created the first periodic table of elements, and with it correctly predicted the properties of elements still to be discovered.

Gerardus Mercator

1512–1594

Flemish

Cartographer

The cartographer Gerardus Mercator invented a system of drawing maps in which the lines of latitude and longitude appear as straight parallel lines, a projection which enabled seaman and navigators to more accurately chart their course. Thus Mercator not only helped to change the world, but also changed the way in which we look at it.

Born in the Flemish town of Rupelmonde on March 5 1512, he was educated at Hertogenbosch in the Netherlands, and later at the University of Leuven where studied he mathematics and became particularly skilled in the making of precise mathematical instruments. It was here that he met the cartographer Gemma Frisius, initially as a student and then as a colleague, and put his engraving skills to good use in assisting in the construction of a terrestrial globe. Drawing on this experience, he began to take a great interest in cartography and produced his own map of Palestine in 1537 and a world map in 1538.

In the 16th century the Netherlands, then part of the Spanish empire, were in a constant state of political and religious upheaval, and Mercator was caught up when he was arrested for heresy in 1544 having shown sympathy for the Protestant cause. However he was released after a few months, but moved to Duisberg in the German Duchy of Cleves in 1552. Here he set up a cartographic establishment, worked as surveyor for the city and taught mathematics at the university. His work brought him to the attention of William, Duke of Cleves, who appointed him Court Cosmographer in 1564.

It was in this period that he invented the system of drawing charts that now bears his name. Europe was expanding rapidly as merchants and sailors crossed the seas and oceans in ever increasing numbers, which stimulated the demand for accurate charts. Although a considerable amount of information was

Above: In 1569 Mercator produced his map, with meridians and parallels of latitude appearing as lines crossing at right angles.

becoming available in respect of the relative position, size and shape of various land masses and their coastlines, it was difficult to present it in such a way as to be useful to sailors and navigators. The problem was the eternal one of portraying an image of the spheroid Earth's surface on a flat sheet of paper. Mercator's projection, as it is termed, showed lines of latitude

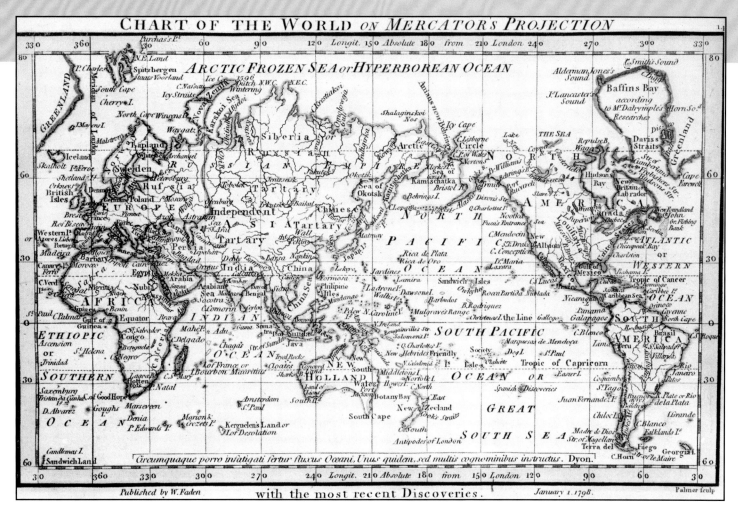

Above: A chart of the world as it was known c.1798 using Mercator's Projection, showing recent discoveries such as New Holland (Australia).

Right: Mercator's cosmic globe with calibrated brass horizon ring and colored drawings of the constellations, c.1551.

and longitude as straight horizontal and vertical lines, and a straight line drawn on such a map would represent an arc of a Great Circle, the shortest distance between two points on the actual surface of the Earth.

The advantages of this system were quickly realized and Mercator charts are still used for navigation today. The maps were also quickly adopted by geographers, publishers and others, and the great majority of printed maps of the world continue use Mercator's projection, despite the challenge from modern alternatives.

Mercator himself worked on in Duisberg producing a great variety of maps, as well some beautifully made globes, celestial and terrestrial, which are much sought-after by collectors today. He died a wealthy and respected man in 1594 and is buried in the city's San Salvatorus cathedral.

Michelangelo Buonarroti

1475–1564

Italian

Renaissance polymath, sculptor, artist and architect

Above: The statue of David originally stood in the Palazzo Vecchio but was moved in 1873 to the Accademia Gallery in Florence.

Michelangelo is the pre-eminent sculptor and painter of the later Renaissance, whose work influenced generations of artists. Only Michelangelo's contemporary, Leonardo da Vinci, could rival his talent. Perhaps Michelangelo's greatest gift was the ability to make marble seemingly come to life, with work such as his magnificent *David* and the heartbreaking *Pietà*. Michelangelo's greatest undertaking, however, was the painting of the Sistine Chapel ceiling in the Vatican, which even his bitterest rivals acknowledged to be a masterpiece of both endurance and talent.

Left: This later portrait of Michelangelo was taken from a self-portrait, but the original is not now believed to be authentic.

Michelangelo de Lodovico Buonarroti Simon was born on March 6 1475 in the small village of Caprese in Tuscany. By the age of 13 Michelangelo already knew he wanted to become a painter, specifically at the workshop of Domenico Ghirlandaio. There his talent was so evident that after only a year he moved to the sculpture school in the Medici gardens where he was introduced to human anatomy and allowed to study corpses.

Soon Michelangelo was invited into the household of Lorenzo de Medici where he met and mingled with many of the great artists, poets and scholars of the Florentine Renaissance. When Lorenzo's death resulted in dangerous political turmoil, Michelangelo escaped to Rome where he admired and studied the magnificent classical sculptures that had only recently been excavated. One of his greatest works, *Pietà*, was carved during this period while he was still under 25 years old. It was placed in Saint Peter's Basilica.

Michelangelo returned home to Florence when the Arte della Lana (Wool Guild) commissioned him to produce a colossal marble statue of *David*. Completed by 1504, arches had to be pulled down and streets widened as 40 men took five days to get it into position in the Palazzo Vecchio.

Michelangelo's reputation as the greatest living sculptor led Pope Julius II to insist he come to Rome to work on his tomb. But once there, Michelangelo was diverted to work on the Sistine Chapel ceiling, even though he was not a painter. He tried to refuse the commission but reluctantly conceded when he was allowed to follow his own composition. The result was four hard years of painting, and well over 300 larger-than-life figures showing nine scenes from the Book of Genesis surrounded by a variety of Old Testament subjects.

Three decades later, in 1534, Michelangelo returned to the Sistine Chapel, this time to work on *The Last Judgment*, painted in fresco behind the altar. Finally finished in 1541, it caused great controversy for its "obscene" nude figures.

At an age when most artists had retired Michelangelo still kept working and was in his 70s when he was made chief architect of St Peter's Basilica; he refused payment for what he considered to be sacred work. It was a demanding job, especially the construction of the dome; furthermore he had to constantly refute his critics, all the while trying to prevent the funds for the project disappearing in bribes and fraudulent practices.

Michelangelo's final masterpiece was the *Rondanini Pietà* for his own tomb, carved when he was in his 90s. In February 1564 Michelangelo caught a fever and died on the 18th, his body was secreted out of Rome and taken back to Florence by his nephew, where it was greeted by grieving crowds before burial in Santa Croce.

Below: Michelangelo worked on the Sistine Chapel ceiling between 1508 and 1512, it shows the Downfall of Man *and the* Promise of Salvation.

Below: Michelangelo's chalk and brush drawing of a large decorative candlestick lay undiscovered for years in the National Galleries of Scotland

Following pages: The enormously ambitious project centers around nine episodes from the Book of Genesis and depicts over 300 figures.

Marilyn Monroe

1928–1962
American
Film actress

More than just a movie star or glamour queen, Marilyn Monroe became a global sensation and a 20th century icon. Her sex appeal, talent and untimely death combined to make her an enduring star, recognizable all over the world. Her image as the glamorous blonde sex symbol started a 'look' that has been copied around the globe, one that in many ways still epitomizes the concept of feminine beauty and allure. Although playing the archetypal 'dumb blonde' in her early films, the sheer power of her personality and unique image made her an international celebrity, and gaining acclaim as a talented actress in later films. Her name remains synonymous with beauty, sensuality and sparkle – and her life an inspiration to all who struggle to overcome personal obstacles for the goal of success.

She was born Norma Jean Mortenson on June 1, 1926 in Los Angeles, California, to Gladys Baker. The identity of her father being uncertain, she was baptized Norma Jean Baker. Gladys was eventually committed to a mental institution and Norma Jean spent most of her childhood in foster homes

and orphanages until 1937. In 1942, aged 16 and unwilling to return to an orphanage, she opted to marry 21-year-old Jimmy Dougherty. When Jimmy became a sailor, Norma Jean took an assembly line job in a munitions factory in Burbank, California. There, photographer David Conover saw her and knew he had found a "photographer's dream." Within two years, she was a reputable model with many popular magazine covers to her credit. Divorcing Jimmy in June 1946, she dyed her hair blonde and changed her name to Marilyn Monroe (borrowing her grandmother's last name) and headed for Hollywood.

Marilyn's first film role was a bit part in 1947, but she worked steadily until her performance in 1953's *Niagara* took her to stardom and to leading roles in hugely popular films like *Gentlemen Prefer Blondes, Bus Stop* and *Some Like It Hot.*

Unfortunately, her personal life was somewhat disorganized. Affairs were rumored with both John and Robert Kennedy and her other husbands included baseball star Joe DiMaggio (1954) and playwright Arthur Miller (1956–61). Her last film, in 1961, was Miller's drama *The Misfits.* In all, she made 30 films, leaving behind the unfinished *Something's got to Give.*

She was found dead on the early morning of August 5, 1962, at her Brentwood, California home. Even death did not end her fame, as conspiracies and alleged involvement by the highest in the land were soon circulating.

As Norma Jean she had a lonely, tragic life. But as Marilyn Monroe, she

Above: By 1952 Monroe was hitting her career peak and actively promoting her hugely popular "dumb blonde" image.

Above: In June 1956 Monroe married Arthur Miller in a secret civil ceremony in White Plains, NY; they divorced in 1961.

Above right: Monroe aged about 22 and at a time when she had only made a few uncredited movies.

Right: Monroe in Let's Make Love, *directed by George Cukor for 20th Century Fox, it was one of her last movies.*

personified Hollywood glamour – an alluring beauty, with voluptuous curves and a generous pout. Yet, Marilyn was more than a sex goddess, as a seeming innocence, combined with innate sensuality, endeared her to the world. Dominating the 1950s movie-star era, she became the most famous woman of the 20th century – despite a legacy of opinions about her life and many questions surrounding her death.

Morita Akio

1921–1999

Japanese

Co-founder of the Sony Corporation.

Morita Akio co-founded the Sony Corporation, the world-renowned company, which introduced many innovative consumer electronics such as the transistor radio, to the mass marketplace in the second half of the 20th century.

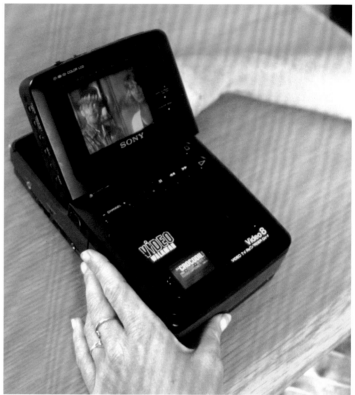

Above: The Sony GV-8 8mm video walkman was introduced in 1986 so viewers can watch 8mm movies or television where ever they are.

Left: Morita Akio co-founded the electronics giant Sony and in the process revolutionized the world of electronics.

Morita was no stranger to the world of business, as he came from a family involved in the production of such products as soy sauce and sake. As the oldest child it was always thought that he would eventually take over the family business, but his leanings were towards a different world, particularly after graduating from Osaka Imperial University with a physics degree in 1944. The Second World War raged at the time and he rose to the level of lieutenant in the navy.

Unperturbed by the destruction and devastation that beset Japan at the time, Morita founded the Tokyo Telecommunications Engineering Corporation in 1946, partnered by Ibuka Masaru, an older navy colleague who would bring to the relationship expertise in the areas of engineering and product design. Morita's own family became the largest shareholders in the new venture.

In a relatively short space of time, the company had

developed magnetic recording tape and sold the very first tape recorder in Japan. By 1957 the revolutionary, fully transistorized, pocket-sized radio had hit the market, driven by Morita's promotion of its portability. The following year the company became the Sony Corporation.

The newly named company developed quickly. The Sony Corporation of America was established in 1960, at the same time as the world marveled at the company's launch of the world's first transistor television. In 1966, Morita authored a book entitled *Never Mind School Records*, which emphasized the individual's natural attributes in becoming a successful businessman, rather than academic prowess.

Sony's star continued in the ascendancy, as the century progressed and in 1979 the Walkman portable music player burst upon the world, to be followed five years later by the compact disc player. Morita was largely responsible for pushing the Walkman brand, which became one of the company's most popular products. After the disaster of the Betamax video recorder format, Morita realized the need for industry standards within electronics, so when Sony developed the CD storage disk, they did so in tandem with the Dutch firm Philips Electronics. By the close of the 1980s Sony owned both

Above: Sony headquarters towers in Tokyo in 2003, when the company paid almost five billion USD to acquire Metro-Goldwyn-Mayer.

Left: Morita with co-founder and partner Ibuka Masaru: they led Sony to top of the Japanese and then the world's electronics industry.

Columbia Records and Columbia Pictures, major symbols of US show-business dominance. In 1982 Morita became the first Japanese national to be awarded the Albert Medal by the UK's Royal Society of Arts, just one of a number of honors bestowed upon him.

Morita suffered a cerebral hemorrhage in 1993 and declining health forced his resignation as chairman of Sony a year later – a bitter blow for a still highly active individual. His role was taken over by Norio Ohga who had once been critical of the company's products. Morita eventually died of pneumonia in October 1999, leaving behind him a global brand which had its roots in a company set up purely and simply to supply a war-ravaged nation.

Samuel Morse

1791–1872
American
Inventor

Above: In May 1844 Samuel Morse sent the first telegraph message using his code, it said, "What hath God wrought."

Samuel Finley Breese Morse, the inventor of the wire telegraph, is best remembered not for his invention, but for the code that bears his name, which he developed to allow messages to be sent quickly and reliably over long distances.

Morse was the son of a pastor, and was born in Charlestown, Massachusetts. He went on to study at Yale, and then in 1811 he traveled to Britain, to study art in London's Royal Academy. By the time he returned to America in 1815, he had become an accomplished and sought-after painter. For the next decade he devoted himself to his painting, and his subjects included President Monroe and the Marquis de Lafayette. In 1818 he married Lucretia Walker, from a prominent New England family, and she bore Samuel three children. However, Samuel spent much of his time in New York, where he maintained a studio.

In 1825 Morse was working in Washington DC when he received a letter that transformed his life. It was from his father, telling Morse that his wife had died. By the time he arrived back in Massachusetts she had already been buried. As a result he vowed to find a way to improve long-distance communication. After her death he spent three years in Europe, and on the sea voyage home he met the Boston scientist Charles T. Jackson, who introduced Morse to the concept of electromagnetism. This chance meeting provided Morse with the opportunity to change the world. His subsequent experiments with electromagnetism resulted in the invention of the single-wire telegraph.

He patented America's first telegraph machine in 1837, by which time other inventors were also developing similar machines. The German inventors Carl Friedrich Gauss and Wilhelm Weber produced an electromagnetic telegraph in 1833, while in 1837 William Fothergill Cooke and Charles Wheatstone patented an electric telegraph in Britain. However, Samuel Morse's design was more efficient and reliable, and his signals could be transmitted for much longer distances.

He toured America and Europe in search of backers, but it was the US government who would first realize the potential of Morse's invention. In 1844 it built an experimental telegraph

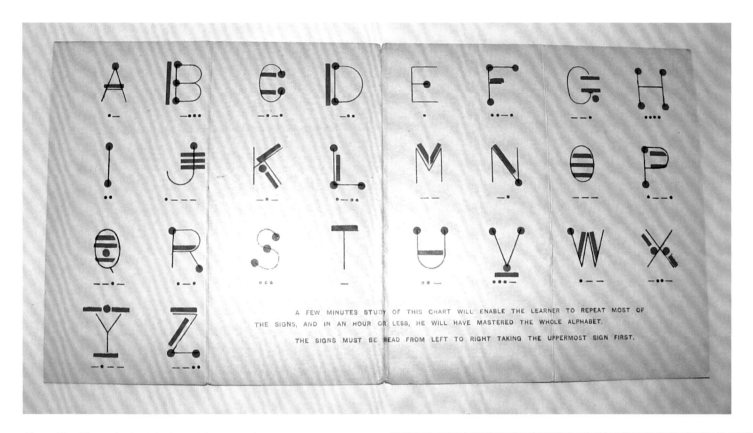

Above: The Morse Code alphabet, with descriptions showing how to read the symbols.

Right: An early 20th centuary photograph of a woman using a morse code radio communications device housed within an oak box.

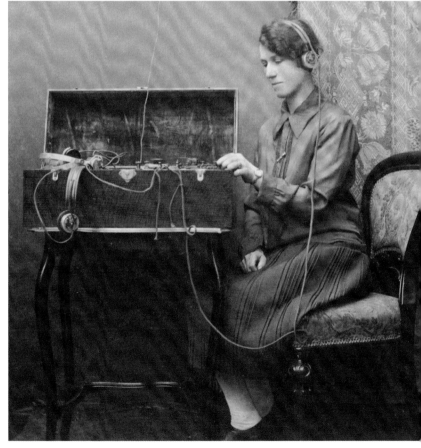

line between Washington DC and Baltimore, Maryland and in May Morse sent its first signal – the message "What hath God wrought", using the special code Morse had devised. By 1845 he had established the Magnetic Telegraph Company, and telegraph lines began to snake their way across America.

Morse also fought lawsuits to defend his patent, and to see off his rivals. As a result he was declared "the inventor of the telegraph". However, despite accolades from other countries, in America he never received official recognition for his invention until the years immediately before his death. However, within a decade his "Morse Code" had become an international system, and the Morse single-wire telegraph system spread its web on every continent. By the time of his death at the age of 80, Morse's invention was in widespread use, and had more than achieved his goal of transforming long-distance land communications.

Wolfgang Amadeus Mozart

1756–1791

Austrian

Composer

One of the most talented musical composers in history, Mozart was a genius who composed in all the musical genres of his time, and did so brilliantly.

Johannes Chrysostom Wolfgang Amadeus Mozart was born in Salzburg, Austria in 1756. His father, Leopold, worked as a composer and music teacher and saw the potential for extraordinary talent in young Wolfgang at a very early age. Mozart was playing the harpsichord by the age of three, composing by the age of five, he produced his first minuet at age six, and began a European tour, (performing with his older sister, Maria Anna, also a child prodigy), when he was seven. The family traveled throughout Mozart's formative years and performed for the courts of Mannheim, Munich, Paris, London and The Hague. Along the way, Mozart wrote his first opera at age eight, his first oratorio when he was 11, and his first symphony by the age of twelve.

In 1781 Mozart moved to Vienna and, having left his family behind, attempted to support himself as a freelance musician. He quickly developed a reputation for being the finest keyboard player in the city. At this time he also fell in love and married Constanze Weber. In Vienna Mozart met the famed composer Joseph Haydn, who would prove to be a major influence on Mozart's work and life, with Mozart referring to him as "Papa Haydn." The two often played in a string quartet and Mozart wrote a series of quartets that he dedicated to his friend, now known as the *Haydn Quartets*. From 1782–1785 Mozart performed many concerts as a soloist where he would frequently premiere new work. He had a ferocious work ethic, often producing three or four new pieces each week. These shows were immensely successful and, with money pouring in, he and his wife began to live a very extravagant lifestyle.

Mozart turned away from his profitable career as a playing musician to concentrate on writing, and in 1786 collaborated with Lorenzo da Ponte on two of his most famous works, the operas *The Marriage of Figaro* and *Don Giovanni*. Despite the critical and popular praise for these new works, Mozart fell on hard times, as his writing career was much less profitable than his income from performing. With Austria at war it was a difficult time for musicians to make a living, and Mozart found himself borrowing money and struggling to pay the bills.

Left: Mozart struggled to make a living after he stopped performing in 1785 in order to write more.

In the final year of his life Mozart released one of his most admired pieces, *The Magic Flute*, and thanks to the patronage of Dutch and Hungarian aristocrats was able to start repaying his debts.

Only 35 years of age when he died, Mozart is arguably the most talented and creative composer in history. In his short lifetime he was able to produce 21 piano concertos, 24 sting quartets, 35 violin sonatas, 5 violin concertos, concertos for clarinet and other wind instruments, chamber music, masses, 22 operas, and more than 45 symphonies. His mastery of all the musical genres, along with his creative imagination and prodigious and prolific talent, mark him as an undisputed genius.

Right: Mozart was a child prodigy and made his first European tour, with his older sister, at the age of only seven.

Below: Thousands gather for celebrations marking the 250th birthday of Wolfgang Amadeus Mozart, in Salzburg on January 27, 2006.

Rupert Murdoch

Born 1931

Australian/American

Media proprietor

Starting with just a small Australian newspaper, Rupert Murdoch has made the giant News Corporation one of the largest and most influential media company in the world. As News Corporation's CEO, Murdoch wields immense power with his global media company. His empire covers television, films, cable networks, book publishing, satellite television, magazines and newspapers operating in the USA, Australia, the UK, Europe, Asia and the Pacific Rim. In effect, it could be said that he runs the world's news machine.

Born Keith Rupert Murdoch on March 11, 1931 in

Below: Murdoch acquired the Sydney Daily Mirror in 1960 and then went on to buy many of the world's popular newspapers.

Melbourne, Victoria, he went on to study in Britain at Oxford University. After a brief spell in London on the *Daily Express*, he returned to Australia to take over from his father, Sir Keith Murdoch, as owner of the *Adelaide News*. Using it as a base, Rupert Murdoch first made the small town newspaper sufficiently profitable for him to move from Adelaide and began his mission to buy companies, expand them or create them.

Gradually, he acquired some of the most popular and widely read newspapers in Australia, the UK and the USA. His rise was undoubtedly the result of a ruthless management style coupled with a desire to succeed. His progress often caused controversy and outrage as he tried to eradicate what he perceived to be corruption, waste and overstaffing in his London-based operations. Confronting the powerful print trade unions, Murdoch's actions caused many protests and petitions against him and his companies – with the satirical weekly magazine *Private Eye* dubbing him "The Dirty Digger" in view of his activities and antipodean origins. Nevertheless, his determination to cut costs and increase profits did

produce the results he wanted, although often at the cost of many jobs. As he himself explained, "I'm a catalyst for change . . . you can't be an outsider and be successful over 30 years without leaving a certain amount of scar tissue around the place."

To expand his US television interests, Rupert Murdoch became an American citizen in 1985. Throughout the 1980s, News Corporation continued to grow rapidly, acquiring interests in newspapers, magazine, book publishing, television stations, film and more. Its rapid growth and expansion into satellite TV also brought massive loans. Then, a downturn in the early 1990s meant that many of his American magazine interests were sold to pay off some of the loans. That slight pause ended rapidly and the mogul continued his acquisition

Above: News Corporation founder Murdoch being interviewed by television anchor Neil Cavuto at the Washington Convention Center April 2, 2009.

Left: Murdoch in 1986, a year earlier he had become a U.S. citizen so he could buy into American television.

and growth, taking over some of the largest and most popular brands across the media industry – including 20th Century Fox, Fox Television, DIRECTV, HarperCollins publishers, Festival Records and the *New York Post*.

The effect that Rupert Murdoch has had on the world's media and entertainment industry is massive. Along the way he created an empire that has gone from strength to strength, securing positions in every important media and country. His great success and power have also created many critics, but Murdoch continues ignore them, seemingly happy with life with his young wife and continuing to grow his global media empire.

Sir Isaac Newton

1643–1727

English

Physicist, mathematician and philosopher

Sir Isaac Newton's soaring intellect effectively set the ground rules for modern science, and his ideas and methods were the foundations upon which those who followed in his footsteps could build to make further great advances.

Born at Woolsthorpe in Lincolnshire, he went up to Cambridge University at the age of 18 in 1661 and was elected a Fellow of Trinity College six years later, becoming Lucasian Professor of Mathematics in 1669. He remained at the university until 1696 and it was during his time at Cambridge that much of his life's work was achieved. So broad was his range of interests and ideas that it is difficult to enumerate them in a few words. Perhaps his greatest bequest to science was the publication of his *Philosophiae Naturalis Principia Mathematica* (Mathematical Principles of Natural Philosophy), more universally known as *Principia*, which was eventually published in 1687. This work comprised three volumes of which the first set out his laws of mechanics including the concept of gravitation. The second covered the theory of fluids, including air, and motion through them, while the third applied the theory of gravitation to the universe and provided a framework for predicting the paths of heavenly bodies. It also established the influence of gravity on the Earth's tides, enabling accurate tables to be constructed.

In another field Newton made significant advances in the science of optics, particularly in the refraction of light and the construction of light patterns of differing colors. This was to have significant implications in the construction of precision optical instruments such as telescopes and microscopes.

Newton's other great invention was more abstract. He was undoubtedly the father of the mathematical process known as calculus, which is absolutely essential in so many scientific and mathematical processes in the modern world. Unfortunately, like many advances of the time, there were others who solved part of the problem and subsequently claimed the whole for

Below: Sir Isaac Newton was knighted by Queen Anne when she visited Cambridge University in 1705.

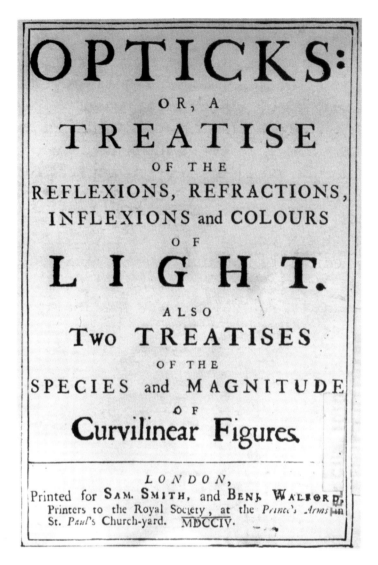

OPTICKS:
OR, A
TREATISE
OF THE
REFLEXIONS, REFRACTIONS,
INFLEXIONS and COLOURS
OF
LIGHT.
ALSO
Two TREATISES
OF THE
SPECIES and MAGNITUDE
OF
Curvilinear Figures.

LONDON,
Printed for SAM. SMITH, and BENJ. WALFORD,
Printers to the Royal Society, at the *Prince's Arms* in
St. *Paul's* Church-yard. MDCCIV.

Above: Title page of "Opticks" by Sir Isaac Newton, "A treatise of the reflexions, refractions, inflexions and colours of light," published 1754.

Right: Newton's reflecting telescope used a curved mirror to reflect and focus the light and was consequently much reduced in length.

was appointed Warden of the Royal Mint in 1696, becoming Master in 1699. He was knighted by Queen Anne when she visited Cambridge University in 1705. As early as 1671 he was a member of the Royal Society and became its most illustrious president in 1703, retaining the post until his death in 1727. During this period he had the satisfaction of seeing his scientific concepts accepted throughout Europe and he was able to concentrate on refining and publishing his ideas, as well as other works covering chemistry, theology and history. On his death he was accorded the honor of a tomb in Westminster Abbey.

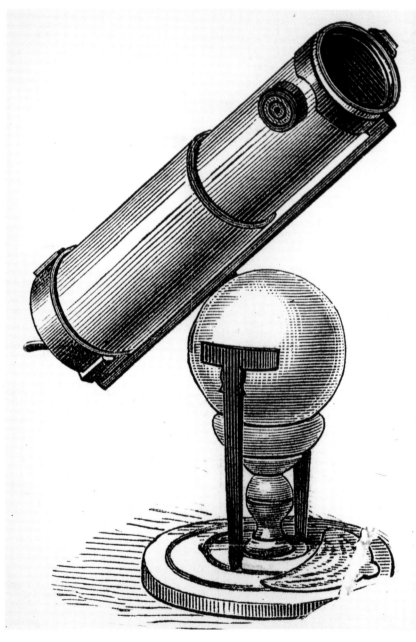

themselves. In particular, Newton became engaged in a bitter feud with the German mathematician Gottfried Leibniz who independently evolved a concept of differential calculus in 1675. However, his work was predated in essence by Newton in 1666 in what he described as his method of fluxions. Nevertheless, the dispute rumbled on, even after Leibniz's death in 1716.

In the meantime, Newton was reaching the peak of his career. He first became a Member of Parliament in 1689 and

Florence Nightingale

1820–1910

English

Nurse and medical reformer

Florence Nightingale, "the Lady with the Lamp" was a pioneering nurse, whose introduction of new systems of hygiene transformed British hospitals in the 19th century.

As a young woman Nightingale turned down proposals of marriage and ignored the social conventions of the day for a woman of her standing. Despite protests from her family, she chose nursing as her profession and in 1850 began her training at the Institute of St. Vincent de Paul in Alexandria, Egypt.

Nightingale continued her studies at the Institute for Protestant Deaconesses near Düsseldorf, Germany for three additional months in 1850. In 1851 she returned to

Above: Within a year of her arrival in the Crimea, Nightingale had transformed the field hospital in Scutari, and the mortality rate had dropped considerably.

Left: Called the "Lady with the Lamp" for her work as a nurse and hospital reformer at Scutari hospital during the Crimean war.

Opposite: Nightingale with a group of nurses from London hospitals at Clandon, her brother-in-law, Sir Harry Verney's home, in 1886.

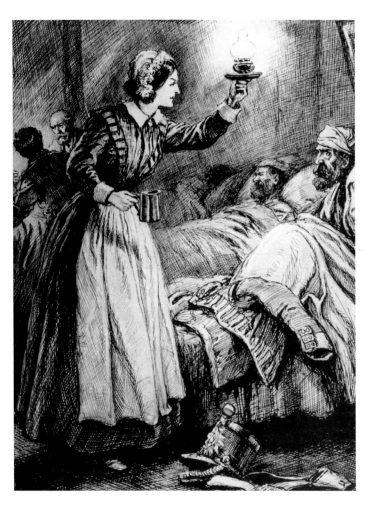

London and accepted an unpaid volunteer position as the Superintendent at the "Establishment for Gentlewomen during Illness". The Crimean War began in March of 1854 and Nightingale was recruited by Sidney Herbert, the British Secretary for War, to become a nursing administrator and to supervise the introduction of a nursing program in military hospitals. Her official title was "Superintendent of the Female Nursing Establishment of the English General Hospitals in Turkey". Nightingale, along with 38 other nurses, was assigned to Scutari, a suburban neighborhood of Constantinople (now Istanbul), in November of 1854.

Upon arrival in Scutari, Nightingale was appalled by the

unsanitary working and living conditions in the hospitals. She found wounded soldiers lying on bare floors surrounded by vermin, with men dying more often from diseases such as cholera and typhus, than from their battle wounds. Though she attempted to improve hospital conditions, Nightingale was met with resistance from military officers and doctors who were offended by her no-nonsense approach. Shunned by her male colleagues, Nightingale took the extraordinary step of using a contact at The Times to write a story about the deplorable conditions of British military hospitals. Soon after the article was published and read by an outraged public, Florence Nightingale was allowed to continue her work, and according to another report in The Times, "she may be observed alone, with a little lamp in her hand, making her solitary rounds." She improved the fresh water supply, set new standards for sanitation, introduced fresh fruit and vegetables into patients'

diets and even used her own money to buy equipment for the hospital. Within one year of arriving in Scutari she had transformed her field hospital and the mortality rate for wounded soldiers dropped from 60% to just 2.2%.

Florence Nightingale returned home to England in 1856. In 1860 she founded the Nightingale Training School for nurses at St Thomas' Hospital in London. Nurses trained under the Nightingale system would soon staff hospitals all over Britain. Unable to continue with the duties of a daily nurse due to an illness she contracted in Scutari, Florence Nightingale went on to publish over 200 works, from books to pamphlets, during the remainder of her life. Two of those books, Notes on Nursing and Notes on Hospitals were enormously influential in shaping our modern health care. Her pioneering efforts to improve sanitation, military health and hospital planning founded practices that are still being used today.

Alfred Nobel

1833–1896

Swedish

Industrialist and Philanthropist

The name of Nobel is inextricably mixed with the prize that bears his name awarded annually to the person or organization that has made a major contribution in activities such as human rights, arms control and prevention or mediation of conflicts around the World. Nobel prizes are also awarded in other field such as Physics, Chemistry, Medicine and Literature, but the Peace Prize is the one that tends to capture public imagination. The paradox in all this is that Alfred Nobel made the fortune, which still finances these awards, from the invention, production and sale of explosives. Although originally developed for civilian use in activities such as mining and quarrying, they inevitably found wide scale military applications.

Born in Stockholm on 21 October 1833, Nobel moved with his family to Saint Petersburg in 1842 and subsequently studied chemistry, both in Russia and later in the United States. He returned home in 1855 and joined the family business, which was involved in the production of explosives, particularly nitroglycerine that was extremely volatile and difficult to handle. This was illustrated only too graphically when Alfred's brother Emil was killed in an accidental explosion at their Heleneborg factory in 1864.

Nobel had already devised a patent detonator to trigger the nitroglycerine, which used a strong percussive shock instead of the traditional heat method. Spurred on by the death of his brother, he sought to find a way of making the explosive safer to handle. The solution was to mix it with a silica-based compound called diatomaceous earth, which allowed the nitroglycerine to be safely handled and molded into convenient shapes. He called his invention Dynamite and it was patented under this name in 1867. Subsequently he

Below: Alfred Nobel's bust presides over the annual Nobel laureate awards ceremony at the concert hall in Stockholm.

international fraternity." The recipients of these prizes were to be determined by Nobel Laureates representing august Swedish institutions, such as the Royal Swedish Academy for Sciences or, in the case of the Peace prize, by a committee made up of five members of the Norwegian parliament. In his usual thorough manner, Nobel detailed all these arrangements in his will which was made in 1895, only a year before his death.

Uniquely Nobel had changed the world on two counts. On the one hand he supplied enough explosives to blow much of it apart, but then balanced this by providing the means by which science could be advanced for the betterment of all mankind.

Above: Swedish chemist and philanthropist Alfred Nobel photographed c.1863. He studied chemistry before joining the family explosives factory at Heleneborg.

Right: Mayor Bloomberg, deputy prime minister of Sweden Winberg, and Crown Prince Haakon of Norway, at the unveiling of the Nobel Monument.

also devised a mixture of gun cotton and nitroglycerine, which was patented as Gelignite (after its jelly like characteristics) in 1876. By then he was the world's foremost authority on explosives and his various companies generated an enormous income. This enabled him to buy up other concerns, notably Bofors – an iron and steel mill – which he transformed into one of the world's leading armament producers.

Alfred Nobel died in 1896 leaving a fortune of over 30 million kronor (worth hundreds of millions US dollars at today's rates). In his will he set aside this money to provide for his eponymous prizes to honor those who excelled in the fields of science, medicine, literature and for "service to the

Barack Obama

Born 1961

American

44th President of the USA

In a cynical age when the words of politicians are often regarded as self-serving, it is refreshing to see a man who not only embodies the hopes of a nation, but also appears to be a charismatic, honest statesman with a genuine desire to improve the world around him. Barack Obama's election as America's first black president in November 2008 was a landmark event, viewed by many as the culmination of years of civil rights protests in the USA.

The son of a Kenyan father and a mother whose family came from Kansas, Barak was born in Hawaii and spent a large part of his childhood in Indonesia. Graduating from Colombia University in New York with a degree in political science in 1983, Obama moved to Chicago where he took a job as a community organizer and helped to establish a job training program, a college preparatory program, and a tenants' rights organization. In 1988 he left Chicago for Harvard Law School, where was elected as an editor for the *Harvard Law Review* in

his first year and president of the journal in his second. He was the first black president and it was this post that first brought him to public attention. While working as a summer intern at the Chicago law firm of Sidley Austin, Obama met Michelle Robinson, a young lawyer who he married in 1992. The couple now has two daughters.

From 1992 to 2004 Obama worked as a professor of constitutional law at the University of Chicago and became a Democratic Party activist, helping to organize Project Vote to register black voters during the Clinton election campaign. In 1996 he ran for his first seat in public office and won a place in the Illinois state senate. As a state senator Obama was successful in gaining bi-partisan support for his legislation reforming health care and ethics laws. Running for election to the US Senate in 2004, he gained national recognition for his keynote address at the 2004 Democratic National Convention. The speech, which drew rave reviews from political pundits and citizens alike, is seen to have struck a helpful and healing tone at a time when the nation was bitterly divided.

Barack Obama announced his candidacy for Presidency of the United States of America on February 10, 2007 only halfway through his first senate term. Obama defeated his rival Senator Hillary Clinton in a hard-fought campaign to win the nomination for the Democratic Party. His speech accepting the nomination was attended by 75,000 people, with approximately 40 million more watching on television. Clearly a serious, committed politician, he has also managed to tap into the times. His campaign was fought over the Internet almost as much as on the traditional hustings, with the slogan "Change we can believe in." His stirring oratory and promises to bring back US troops from Iraq, to close the controversial detention centre at Guantanamo Bay and to restructure the tax system in favor of middle and low-income earners struck a chord with voters

He was elected president in November 2008 with 53% of the vote to become the nation's first African-American President.

Right: The Obama family at the annual White House Easter Egg Roll in April 2009. The ceremony dates back to 1878.

Left: Two days before becoming president, Obama spoke to the "We Are One: The Obama Inaugural Celebration At The Lincoln Memorial."

Above: Obama and Secretary of the Treasury Geithner and Federal Reserve Chairman Bernanke in the Roosevelt Room of the White House.

Following pages: Barack Obama being sworn in by Chief Justice John Roberts as the 44th president of the United States, on the West Front of the Capitol January 20, 2009.

Robert Oppenheimer

1904–1967

American

Physicist

While being dubbed "the father of the atomic bomb" might be seen as a dubious honor, Professor Oppenheimer was also one of the most gifted scientists of his generation, and an advocate of the atom bomb as a tool for peace, rather than a weapon of Armageddon. He also struggled with the moral dilemma where his conscience was at odds with his desire to further the best interests of his country.

Oppenheimer was born into a New York Jewish family, and soon proved a gifted linguist, and a distinguished scholar. After graduating from Harvard University he traveled to Europe,

Above: Oppenheimer and Major General Groves beside the remains of the holding tower after the atomic bomb test, Los Alamos, California.

Left: After the war Oppenheimer advocated the free exchange of scientific knowledge, regardless of political boundaries.

where he studied at the universities of Cambridge, Leiden and Göttingen, where he was awarded his PhD. It was clear that he had an exceptional mind, and by the time he returned to America to teach at Berkeley in 1927, he was widely regarded as one of the leading research physicists of his generation.

However, unlike many scientists, Oppenheimer followed current events, and involved himself in political debate. He was alarmed by the rise of Fascism in Europe, and in 1939 he was greatly concerned when his friend Niels Bohr told him that German scientists were close to splitting the atom, which in turn raised the possibility that the Nazis might be able to

Above: Top secret photograph of an atomic bomb of the "Little Boy" type which was detonated over Hiroshima in 1945.

Right: The deadly mushroom cloud rising above Nagasaki on August 9, 1945 after the second atomic bomb was dropped on Japan

develop nuclear weapons. This prompted President Eisenhower to establish the Manhattan Project in 1941 – America's scientific quest to develop a nuclear bomb. In June 1942 Oppenheimer was appointed its director, and under his guidance a team of leading physicists were recruited, and based in the newly built research station at Los Alamos, New Mexico. Over the next three years Oppenheimer and his team developed the atomic bomb. After its first test explosion in the desert of New Mexico in July 1945, Oppenheimer commented, "We knew the world would not be the same". Two months later, two bombs were detonated over the Japanese cities of Hiroshima and Nagasaki, killing an estimate 100,000 Japanese, most of whom were civilians. With the hindsight of history, this probably saved many more lives, as within a week the Japanese surrendered, so ending the Second World War, and with it the specter of a costly US invasion of the Japanese homeland.

After the war, Oppenheimer chaired the US Atomic Energy Commission, but his opposition to the development of a hydrogen bomb led to his fall from grace. In 1953, during the McCarthy "witch hunts" he was accused of having Communist sympathies, and his security clearance was revoked. Following his shabby treatment at the hands of his government, Oppenheimer devoted what remained of his life to teaching, writing and academic research, serving as the director of Princeton's Institute for Advanced Research. In 1962 he was rehabilitated by President Kennedy, finally receiving the official recognition he deserved. Until his death he remained an advocate of the free exchange of scientific knowledge, regardless of political boundaries. However, despite his achievements in the fields of black hole research, quantum physics and the electron-positron theory, he remains the man who created a weapon with the potential to destroy mankind, rather than a physicist whose abilities opened up new scientific frontiers.

Jesse Owens

1913–1980
American
Athlete

Jesse Owens was a remarkable American athlete who won a record four gold medals at the 1936 Berlin Olympics. His success as a black athlete in the heart of Nazi Berlin was a blow to Adolf Hitler, who wanted to use the games as a showcase

Above: Jesse Owens and Ethiopian marathon runner Abebe Bikila watching the arrival of the Olympic flame for the 1972 Munich Games.

Left: Owens established six world records in 1935. His 100m record stood for 20 years and his long jump for 25 years.

Opposite: August 4, 1936, medal winners for the 100 meters: Martinus Osendarp, Holland, bronze; Owens, gold; and Ralph Metcalf (both United States), silver.

for supposed Aryan supremacy.

James Cleveland Owens was born in Alabama but moved with his family to the Granville neighborhood of Cleveland when he was nine years old. In later life he gave much credit to Charles Riley, his high school coach, who encouraged him and made allowances for his difficulty in making evening training sessions because he understood that Jesse had to work.

Jesse Owens rose to national prominence in 1933, when he equaled the world record (9.4 seconds) for the 100-yard dash. He attended Ohio State University, but without a scholarship he had to continue working part time. At the time, America was a highly segregated society and when traveling with the team, Jesse suffered the indignities of eating at separate restaurants and staying at different hotels from his white teammates.

Jesse Owens' finest moment came in the 1936 Berlin Olympics. He won Olympic gold in the 100 meters, long jump, 200 meters and 4x100 meters relay. It was a powerful rebuttal to the Nazis' hopes of displaying "Aryan superiority." Hitler gave medals to German athletes on the first day, but after that decided not to award any more medals in person. Albert Speer later wrote that Hitler was annoyed that the "negro, Jesse Owens had won so many gold medals." Yet, with great irony, Jesse Owens was treated well during his stay in Germany, as he didn't experience the segregation that he did back home in the United States and many Germans sought his autograph.

Despite his great athletic achievement, Jesse Owens was denied the commercial success that was shared by his Caucasian counterparts. Jesse was forced to take part in various "athletic showcases" such as racing against horses, or racing against local runners with a 10-yard head start. As Jesse Owens wryly remarked, "After I came home from the 1936 Olympics with my four medals, it became increasingly apparent that everyone was going to slap me on the back, want to shake my hand or have me up to their suite. But no one was going to offer me a job."

He moved into business but it was not successful and it ended in bankruptcy in the 1960s. However, in 1966, with the civil rights movement gaining impetus, Jesse Owens was given the opportunity to act as a goodwill ambassador, speaking to large corporations and the Olympic movement. Jesse Owens was inducted to the Alabama Sports Hall of Fame in 1970. He was awarded the Presidential Medal of Freedom in 1976 by Gerald Ford and (posthumously) the Congressional Gold Medal by George H. W. Bush on March 28, 1990. His birthplace in Oakville, Alabama was dedicated a park in his honor at the same time that the Olympic Torch came through the community in 2001, 60 years after his Olympic triumph.

Kerry Packer

1937–2005

Australian

Media proprietor

The uncompromising Kerry Packer was an extremely influential television and magazine proprietor skilled in political lobbying. Although an enthusiastic gambler, he became Australia's richest man with a fortune once estimated at A$7 billion. However, in some ways, he is best known for the sporting revolution he started in 1977 with his rebel World Series Cricket –dragging cricket into the television age, changing the sport forever and its image on the global stage.

Kerry Francis Bullmore Packer was born on December 17, 1937 and grew up in the shadow of his father, Sir Frank Packer, the rambunctious Sydney media mogul and a man who always treated his younger son with a chilling lack of consideration. Yet, Kerry Packer was to outstrip his father in wealth, fame and influence, achieving a stature far beyond that of Sir Frank or others of the Packer media dynasty that dated from the turn of the 19th century with Kerry's grandfather, the Tasmanian-born Robert Clyde Packer.

When Sir Frank died in 1974, Kerry at 37 became chairman of the Packer empire of two TV and five radio stations, nine provincial papers and Australia's biggest magazine publisher, plus property and other interests. Under his father's ownership, the hard-hitting *Sydney Telegraph* had moved steadily to the right, giving working-class readers both entertainment and clear policies, and the most famous magazine in Australia, *The Australian Women's*

Above: Packer died age 68 and was mourned with a minute's silence at the Melbourne Cricket Ground in 2005.

Weekly had been linked with the *Telegraph* under Packer's Consolidated Press. The TV stations were there because Frank Packer had been a pioneer of television when it reached Australia in 1956, setting up Channel Nine in Sydney.

Like his grandfather and father, Kerry Packer had a reputation as a bully to his employees, especially his executives, and presented a profanity-rich manner to politicians and businessmen. Yet, he could sometimes show rough-hewn charm, with long-serving staff, valued friends and good causes benefitting from his extraordinary (sometimes anonymous) generosity. He cultivated only those politicians who could be of use and, ironically, had little time for journalists.

Packer collapsed with a heart attack while playing polo in October 1990. Technically, he was dead for eight minutes and when revived by ambulance officers commented laconically, "I've seen what's on the other side, and believe me, there's nothing there." A quintuple heart bypass followed, but even the ailing Packer continued to take a vigorous interest in business – such as Melbourne's Crown Casino, petrochemicals, heavy engineering, ski resorts, rural properties, diamond exploration, coal mines, and supermarket coupons.

Overall, for many cricket traditionalists, Kerry Packer still remains the villain who destroyed a comfortable, amateur-run game in the interests of television ratings and sheer profit. The reality is more that he ensured players were properly paid for the first time, forcing cricket to face commercial realities. His radical changes are even more evident on our screens today as the game expands and evolves with huge payments and show business razzamatazz. Appointed a Companion of the Order of Australia in 1983, he died on December 26, 2005, aged 68 and was mourned with a minute's silence at the Melbourne Cricket Ground.

Above: Packer was a wealthy Australian entrepreneur with many and varied business interests in sport, petrochemicals, mining, and resorts.

Below: World Series Cricket Supertest Grand Final match between WSC Australia and WSC World XI held at the Sydney Cricket Ground. Packer's legacy was to make cricket a more commercial affair.

Emmeline Pankhurst

1858–1928

British

Suffragette and political activist

Above: Emmeline Pankhurst and her daughter Christabel wearing prison uniforms during a spell in jail for demonstrating for women's rights.

Left: When World War I started 1914 Pankhurst halted her militant campaigning and became an advocate of women's involvement in the war effort.

While roundly criticized for her publicity stunts and militant tactics, Pankhurst was nevertheless highly successful in her suffragette campaign in Britain, and in the establishment of basic rights for British women.

Emmeline Goulden had an unremarkable upbringing in Manchester, and after marrying the barrister Richard Pankhurst she played the role of the dutiful Victorian wife, giving birth to five children in ten years. However, Richard was also a vocal supporter of women's right to vote, and Emmeline (or "Emily") became involved in the Women's Franchise League, which advocated women's suffrage. When her husband died in 1898 she took employment as a local registrar of Births, Marriages and Deaths, a post which gave her an insight into the plight of women less fortunate than herself. In the process

she realized that the status quo needed to be shaken up in order to encourage social change.

Frustrated by the lack of progress in women's suffrage, in 1903 she founded the Women's Social and Political Union, a suffrage movement where, as she put it, "Deeds, not words was to be our permanent motto". As such it stood apart from other political movements, and soon developed a reputation for its militancy. In 1908 Emily was arrested when trying to force her way into the House of Commons, to petition the Prime Minister, Lord Asquith. She spent six weeks in prison, and used her incarceration as a means of gaining publicity for her cause. She was arrested another seven times during her political career, each time as a result of a carefully planned publicity stunt. Meanwhile, her campaign was growing in popularity, and a WSPU rally in Hyde Park attracted almost half a million supporters.

The government remained indifferent, so Pankhurst resorted to increasingly militant tactics, despite the rifts this caused within the suffrage movement. Women prisoners went on hunger strike, demonstrations became increasingly violent, and suffragettes even used arson as a means of drawing publicity on their cause. On the outbreak of the First World War in 1914 Emily halted its militant campaign until the end of the war, and instead she became a passionate advocate of women's involvement in the war effort. In 1918 the British government granted limited women's suffrage – the right of women over the age of 30 to vote. The Representation of the People Act represented a personal triumph for Emily – the result of two decades of campaigning, regardless of the personal cost to her and her family.

By then she had transformed the WSPU into the Women's Party, and she continued to campaign for equal opportunities for women until her death a decade after her dream of women's suffrage was realized. It was only after her death that her true contribution to the political landscape was properly acknowledged. *The Herald Tribune* called her "the most remarkable political and social agitator of the early part of the 20th century, and the supreme protagonist of the campaign for the electoral enfranchisement of women." Few would disagree – what is seen today as a basic political right was only won through the tireless campaigning of this remarkable woman.

Below: Arrested at a demonstration outside Buckingham Palace, London, Pankhurst was frequently criticized for her militant tactics. She spent most of her adult life focused on winning women the right to vote

Rosa Parks

1913–2005

American

Civil rights activist

Above: In 1999 **Time** magazine named Rosa Parks as one of the 20 most influential figures of the 20th century. Her actions in Montgomery, Alabama in 1955 helped to jump-start the civil rights campaign in the USA.

In the first half of the 20th century black and white people were segregated in virtually every aspect of daily life in the southern United States by the "Jim Crow" laws. Train and bus companies were not required to provide separate vehicles for the different races, but they did adopt a seating policy that established separate sections for blacks and whites. School bus transportation was unavailable in any form for black schoolchildren in the South. Rosa Parks remembered attending elementary school in Pine Level, where school buses took white students to their new school and black students had to walk to theirs. "I'd see the bus pass every day . . . But to me, that was a way of life; we had no choice but to accept what was the custom. The bus was among the first ways I realized there was a black world and a white world."

Rosa Louise McCauley was born in Tuskegee, Alabama in 1913. In her adult life she would later move to Montgomery, Alabama where married Raymond Parks, took a job as a seamstress at the Montgomery Fair department store, and became involved with the local branch of the National Association for the Advancement of Colored People (NAACP). On December 1, 1955 Rosa Parks was riding the Cleveland Avenue bus, sitting in the appropriate "colored section," when the bus driver, James Blake, asked Parks and three others to vacate their seats in order to make room for white passengers who were boarding. Three of those people complied with Blake's request but Rosa Parks refused. Blake called the police and had Parks arrested. This simple act of civil disobedience by the 42-year-old Parks, a woman simply trying to get home after a day's work, galvanized the local black community and spearheaded the Montgomery Bus Boycott.

Above: Hundreds of people attended the funeral of Rosa Parks, November 2, 2005, at the Greater Grace Temple in Detroit, Michigan.

Below: Parks displays her Congressional Gold Medal of Honor (awarded by President Clinton in 1999) with U.S. Vice President Al Gore.

Tried and convicted of disorderly conduct and violating a local segregation ordinance, Parks later appealed her conviction and challenged the legality of racial segregation. In 1992 Parks told Lyn Neary of National Public Radio, "I did not want to be mistreated, I did not want to be deprived of a seat that I had paid for. It was just time . . . there was opportunity for me to take a stand to express the way I felt about being treated in that manner."

Her defiance galvanized the civil rights movement and Parks went on to work with Martin Luther King, helping him to acquire national prominence for their campaign. *Time* Magazine voted Rosa Parks one of the 100 Most Influential People of the 20th Century. She was awarded the Medal of Freedom, the highest award given to a civilian citizen, by President Bill Clinton in 1996. When she died on October 24, 2005 she was the first woman to lie in honor in the US Capital rotunda, where her casket was visited by over 50,000 people and her funeral was nationally televised. She is widely heralded as the "Mother of the Civil Rights Movement."

Louis Pasteur

1822–1895

French

Chemist

Above: Germ theory – that microorganisms cause many diseases – was controversial in the 19th century, but is a cornerstone of modern medicine today.

The microbiologist who identified the importance of germs in the spread of disease, Louis Pasteur probably did more to improve the world he lived in than most men of his century. The son of a college headmaster, Pasteur was brought up in Dole, near the Swiss border. He studied at the prestigious École Normale Supérieure in Paris, and by the time he was 26 he had become the Professor of Chemistry at the University of Strasbourg. It was there that he courted and married Marie Laurent, the daughter of the university's rector.

His early research was as a chemist, studying the crystalline qualities of tartaric acid. This led to the discovery of molecular chirality – the fact that not all molecules could be superimposed on their mirror image. His example was the difference between gloves for the right and left hands – although they looked similar, a left-handed glove couldn't be worn on the other hand. This research set the groundwork for subsequent breakthroughs in the field of molecular chemistry – research that ultimately resulted in the identification of DNA.

However, by 1854, when Pasteur took up a new professorship at the University of Lille, his principal research was in microbiology. More specifically, he became fascinated by the growth of microorganisms. He proved they developed through biogenesis (i.e. in a natural biological way), thereby disproving the long-held theory of spontaneous generation – the creation of life from inanimate matter. This led to his development of germ theory – the notion that microorganisms are the cause of many diseases, attacking the body from outside. Although this pathogenic theory of medicine was controversial at the time, it is now seen as a cornerstone of modern medicine. While Pasteur was not the first scientist to propose the germ theory, he was the first to prove its existence, and he was therefore able to convince the world of its importance. This alone earned him the modern appellation of the "father of bacteriology," and the pioneer of the medical fight against disease.

A byproduct of this bacteriological research was "pasteurization" – the process whereby liquids such as milk could be heated to kill harmful bacteria. This came about through his study of the way bacteria was responsible for the souring of beer and wine. Together with fellow microbiologist Claude Bernard, Pasteur successfully demonstrated the benefits of pasteurization in 1862.

His work eventually involved research into the causes of cholera, tuberculosis, smallpox and rabies. In the process he pioneered the development of immunization through vaccination – injecting the patient with a small dose of harmful bacteria, to stimulate the body's natural defenses against it. His development of a successful rabies vaccination represented one of the great medical breakthroughs of his age. It proved that the same principle could be applied to other life-threatening diseases. In 1888 the Pasteur Institute was founded in Paris, to develop new treatment against disease, and Louis Pasteur remained its director until his death seven years later. While he died a French national hero, his greatest achievement was the saving of life, both through his own research, and the work of the institute he helped create.

Right: Pasteur at work with an early sterilizer at the Pasteur Institute in 1870. His technique of heating liquids to a precise temperature to kill bacteria became known as pasteurization.

Below: Pasteur also developed a successful vaccine to treat rabies and vaccination techniques to control bacterial infection.

Pablo Picasso

1881–1973

Spanish

Artist and sculptor

Pablo Picasso was one of the most significant and innovative artists of the 20th century. Although he painted in a wide range of styles, he is best remembered for co-founding (with Georges Braque) Cubism, a school of art where objects are broken up and reassembled in abstracted form and seen from a number of angles.

Born in Malaga, Spain, Picasso showed an early affinity for drawing and received training in formal art from his father, an artist and professor. He was admitted to the School of Fine Arts in Barcelona aged 13. At 16, he enrolled at Madrid's Royal Academy, the leading art school in Spain but his lack of discipline and problems with formal tuition made this a brief

Above: Picasso smoking a filtered cigarette in front of his 1917 painting of his first wife, Russian ballet dancer Olga Kokhlova.

Left: **Portrait of Dora Maar** (1937). Born Henriette Theodora Markovitch, she was Picasso's muse and lover for almost nine years.

sojourn. In 1900, he traveled to Paris, then Europe's capital of art, setting up his own studio in Montmartre the following year. Absorbing the influences of the neo-impressionists, he painted *Longchamps* and *The Blue Room* in 1901, marking the start of his Blue Period, which referred to both color and mood. This contrasts with the livelier Rose (or Pink) Period from 1905–1907. In 1907 he produced one of his most famous works, the large and fractured image *Les Demoiselles d'Avignon*.

In tandem with French painter Georges Braque, Picasso pioneered Cubism, a revolution in both painting and sculpture, which also influenced literature and music. A radical, avant-garde movement, Cubism is usually divided into two periods. Analytic Cubism saw Picasso trying to render three-

dimensional objects on flat canvas without resorting to the use of perspective. In Synthetic Cubism, newspaper clippings and other materials were used as part of the painting, creating the first collages in fine art.

After the First World War, Picasso returned to a neo-classical style of painting and, from 1917, worked with Sergei Diaghilev's Ballets Russes, designing costumes and sets in both neo-classicist and Cubist styles, helping the latter to become more publicly acceptable. In 1937, he produced *Guernica*, arguably his best-known work. Its huge, expressive canvas reflected Picasso's horror at the bombing of the Basque town during the Spanish Civil War.

In addition to painting, Picasso was a prolific sculptor, one of his most famous and controversial designs – which could be a woman, a horse or a bird or a completely abstract shape – has stood in downtown Chicago since 1967. His reputation

Above: In his home studio in his villa "La Californie" in Cannes, Picasso is surrounded by his paintings and sculptures.

as a sculptor only really emerged after his death because he retained so much of his work in his own collection.

His paintings are amongst the most valuable in the world; *Garçon à la Pipe* sold for a record $104 million in 2004. Picasso's influence on art is incalculable. Cubism took painting further from the traditional realistic style than ever before, opening up new avenues for self-expression. Explaining his approach, he said in 1935 that, "Art is not the application of a canon of beauty but what the instinct and the brain can conceive beyond any canon. When we love a woman, we don't start measuring her limbs." Picasso was one of the most versatile artists in history and no artist who came after him could escape his influence.

Gregory Pincus, Min Chueh Chang, John Rock

Gregory Pincus

1903–1967

American

Endocrinologist

Min Chueh Chang

1908–1991

American

Reproductive biologist

John Rock

1890–1984

American

Gynecologist

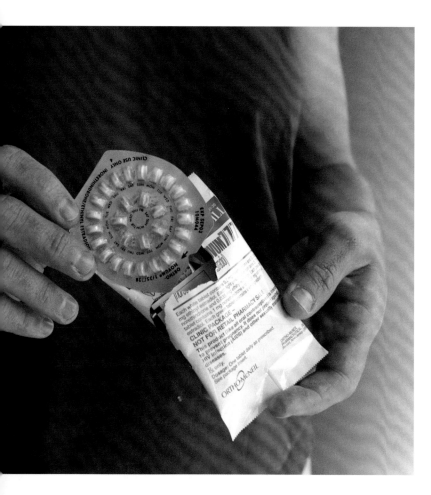

The contraceptive pill has been described as the most significant medical advance of the 20th century, and acclaimed as playing a major role in changing for the better the lives of women worldwide. It has, by extension, also been linked socially to both the women's liberation movement and to the greater sexual freedom associated particularly with the "swinging 60s." Simultaneously, health scares and sociological concerns about its impact have caused controversy.

Explained in the simplest way, the pill works by suppressing ovulation and synthetically mimics the female body's natural hormone cyclical levels of the hormones estrogen and progestin.

Gregory Pincus, a graduate of Cornell and Harvard universities, founded the Worcester Foundation for

Above: Dr Gregory Pincus, among the many honors he received during his lifetime was membership in the National Academy of Sciences of the U.S.A.

Left: A modern blister pack of contraceptive pills, since they were first released in the 1960s usage has soared.

Experimental Biology in 1944, which became a leading centre for the study of mammalian reproduction and the use of steroid hormones. Encouraged by Margaret Sanger, the founder of the US birth control movement, Pincus and his colleagues studied synthesized hormones and their effect on human fertility, discovering that the most effective way to prevent pregnancy in mammals was to inhibit ovulation. In early 1951 Pincus obtained a grant from the Planned Parenthood Association to begin his hormonal contraceptive research alongside reproductive physiologist Min Chueh Chang.

Chang was a Chinese-born reproductive biologist who had studied for his doctorate at Cambridge University and began word at the Worcester Foundation in 1945, to research *in vitro* fertilization techniques. Chang's work testing various steroids and their effect on human fertility when administered orally perfectly complemented Pincus's research. Although credited as co-inventor of the contraceptive pill, Chang spent most of his career working on *in vitro* fertilization, the process that eventually produced the first "test tube baby."

With no serious research funding, Sanger wrote a brief note mentioning Pincus' research to her longtime supporter, the biologist Katharine Dexter McCormick. Together, they enlisted the gynecologist John Rock to begin clinical trials. Rock, incidentally, was a Roman Catholic, who fervently believed that the contraceptive pill should be regarded as a natural method of birth control. He had taught obstetrics and gynecology at Harvard for 30 years and had established the country's first birth control clinic to educate Catholic couples in calendar-based methods of birth control. Rock also pioneered methods of freezing sperm cells and *in vitro* fertilization.

Chang screened nearly 200 chemical compounds in animals and found the three most promising. Then, in December 1954, Rock began the first studies of ovulation suppression potential of the three oral combinations. What their combined work produced was what became universally known in the English language simply as "the pill." Trials were begun in Puerto Rico in 1955 to test the effectiveness of the various synthetic hormones and to note the side effects, and in 1960 the combined oral contraceptive was approved for use in the USA.

Right: Dr John Rock, who, as well his clinical work, wrote a book in 1963 entitled ' The Time Has Come: A Catholic Doctor's Proposals to End the Battle Over Birth Control'

Plato

c. 427–348 BC

Greek

Philosopher

Plato was the founder of modern philosophy and a prolific writer whose theories on knowledge, religion, politics and the state shaped Western thought for over a thousand year and still exert a profound influence today.

Above: Plato lived during the fall of the Athenian empire and is one of the most important philosophers in history.

Left: A Persian miniature shows that Plato's skill with an instrument was such that he could even calm wild animals.

Born into a wealthy family in the city-state of Athens in ancient Greece, Plato's early years are shrouded in mystery. He lived at the time of the fall of the Athenian empire and the subsequent building of another under Philip II of Macedon, father of the all-conquering Alexander the Great.

Plato was a student of Socrates and taught Aristotle. Through Plato, Socrates' teachings were preserved in the form of dialogues (similar to plays) in which Socrates is the main character. Unlike his teacher, Plato left a substantial body of writing, ensuring the survival of his theories, although doubts exist in scholarly circles as to the authenticity of all the works usually attributed to him. He was the first philosopher to satisfactorily formulate the principles of ethics.

The "Theory of Forms" has had a crucial effect on the development of Western religion and philosophy. In it, Plato drew a distinction between knowledge and opinions, and divided the universe into the realm of the visible and that which is beyond the senses; the latter was to form the basis of the Christian concept of heaven. He originated the notion that man is composed of an immortal soul trapped in a mortal body and that the two are separate entities; he theorized that the soul rose up after the earthly existence was terminated, ideas which were later developed by Christian theologians.

Greece, in Plato's time, consisted of independent city-states, continuously at war. Plato disliked battle and the concepts of glory and honor which were associated with it. He directed his mind to conceiving a system of government where people could leave in peace and harmony. He was equally concerned with opposing the Sophists, a group of Athenian philosophers-for-hire who espoused success over morality.

Written in around 380 BC in the form of a dialogue, *The Republic* is Plato's most famous work. It contains the first recorded reference to the concept of utopia and remains one of the foundations of political philosophy. Plato proposes a rigid class structure ruled by the Guardians, philosophers specifically trained in the art of government. The remainder of society is split into two classes – the soldiers and the workers. He went on to discuss justice, education, to which he attached the utmost importance, and property, which, he argued, should be communally owned. Frequently misunderstood and misinterpreted, *The Republic* had a lasting effect on European political thought.

Above: Raphael's fresco version of **The School of Athens**, *where Aristotle and Plato would discuss philosophical problems as they strolled around.*

After the trial and execution of Socrates in 399, Plato fled Athens and traveled in Greece, Egypt and southern Italy. On returning around 387, he founded the Academy, which stood for 1,000 years and was the first institution of learning in the Western world.

Marco Polo

c. 1254–1324

Italian

Explorer

Above: Without a spell in prison as a captive of the Genoese, the story of Marco Polo would have been lost.

Marco Polo was a Venetian merchant and one of the first Europeans to visit China. He was certainly the first whose travels are recorded and the story of his travels, known in English as *The Travels of Marco Polo*, became an important source of European knowledge about China.

Marco Polo was born in about 1254 into a wealthy Venetian merchant family who traded with the East. When he was about six his father Niccolò and uncle Maffeo left Venice for an extended trading expedition eastwards to the Black Sea and then along the Silk Road into central Asia where they joined a diplomatic mission journeying to the fabled court of Kublai Khan, the emperor of China (known at the time as Cathay) in Beijing (then called Dadu). When they left, the great Mongol emperor asked the Polo brothers to ask the pope to send scholars back to China to explain Christianity to him. He also gave them a golden tablet giving them free lodging, horses and food throughout his domains. In about 1269 they arrived back in Venice.

Two years later Pope Gregory X sent out the return expedition: this time the two brothers were accompanied by two Dominican monks (who went home halfway) and the 17-year old Marco Polo. It took them three years journeying through the Holy Land, Persia and central Asia to reach the court of Kublai Khan. They stayed for 17 years, during which time they learnt to speak the language and were given court appointments and gifts.

The emperor particularly favored Marco and sent him as his envoy around China and into Asia into lands where no European had gone before. For three years Marco was even the governor of Yangzhou, a large trading city.

Around 1292, the khan's daughter, Princess Cocachin, was betrothed to Arghun Khan in Persia and the Polos offered to escort her to her husband. The wedding party of 14 ships set out

Above: A beautifully illustrated map showing Marco Polo's long route down the fabled Silk Road to China.

*Below: A 14th century manuscript **Book of the Grand Khan** shows Marco Polo setting off for the Far East.*

from the southern port of Quanzhou and sailed past Borneo, Sumatra and Sri Lanka to southern India and the Persian Gulf. Having safely delivered Cocachin, and loaded with magnificent gifts, the Polos journeyed overland to Constantinople where they took ship for Venice, which they reached in 1295.

The world would not have heard anything more of Marco Polo but in 1298 Venice and Genoa were at war over trading rights and Marco was captured and taken prisoner by the Genoese, probably following the battle of Curzola. Imprisoned for a year he talked about his adventures in China and his travels across Asia. These stories were written down by Rustichelloda Pisa and appeared in book form as *Il Milione*, ("The Million", but more familiarly in English, *The Travels of Marco Polo*.) Scholars are divided as to the exact veracity of his tales, but it is generally accepted that the reports of what he saw are accurate; what he heard, however, is likely to have been exaggerated. The wealth of geographic information detailed by Polo provided a great deal of the information used by the great European explorers of the 15th and 16th centuries.

Elvis Aaron Presley

1935–1977

American

Musician and rock'n'roll pioneer

Elvis Presley, a poor man's son from Tupelo, Mississippi, brought together the threads of white country and black blues to create what we now know as rock'n'roll. And in the eyes of millions he is still The King.

Elvis made his first public performance at 10, winning

Below: A scene from the documentary, **That's The Way It Is** *(1970) which recorded Elvis's remarkable return to live performance in Las Vegas.*

second prize at the Mississippi-Alabama Fair and Dairy show with a rendition of "Old Shep." His appetite for music was huge: he listened to black bluesmen as well as country artists, and it was these strands he would unite and mix with a dash of *Rebel without a Cause* film star James Dean to produce musical dynamite.

Leaving school in 1953 to drive a truck, Presley paid a visit to Sun Records' Memphis Recording Services studios to cut a one-off disc for mother Gladys's birthday. Presley signed to Sun in July 1954 – and, though he did not see out his three-year deal, being transferred to the mighty RCA Victor for a then-enormous $35,000 fee in late 1955, he soon began to make priceless music.

Recording with two older musicians, guitarist Scotty Moore and upright bass-player Bill Black, Elvis scrubbed at his acoustic guitar while looking for his own style. He found it almost by accident when the musicians started casually jamming on an Arthur Crudup blues, "That's All Right." Label boss Sam Phillips told them to do it again with the tapes rolling, and rock'n'roll was born.

New label RCA promoted Elvis's music first nationally and then internationally. TV appearances (in some of which he was cut off at the waist due to his 'lascivious hip-swiveling'), a movie debut in *Love Me Tender*, even being drafted into the US Army in March 1958 all added to the legend, as manager Colonel Tom Parker pulled the strings. In Britain, stars from Marty Wilde to Billy Fury attempted to clone his appeal, while American Elvis followers included Ricky Nelson, Frankie Avalon and Fabian.

Above: Colonel Tom Parker was Elvis's manager right from the start of his career. He steered and manipulated every business decision for the singer.

Below: GI Presley in Fort Chaffee at the beginning of his military service; he spent 17 months in Friedberg, Germany.

While public appearances were put on hold for seven years from 1961 as movies and married life took precedence, the face and music of Elvis Presley continued to exert an appeal that, for many millions of fans, withstood Beatlemania, Motown and psychedelia. His record in the US alone includes 18 chart-topping singles in his lifetime including 'Heartbreak Hotel', 'Hound Dog' and 'Suspicious Minds', and nine Number 1 albums.

Elvis left Hollywood behind and re-launched his music career with the NBC TV show 1968 Comeback Special; this was followed by a return to live performance. But the collapse of his marriage to one-time child bride Priscilla in late 1971 affected him badly and his increasing use of prescription drugs presaged his sad end on 16 August 1977 when he died of heart failure at his palatial Graceland home.

Elvis was the first and greatest star of the rock'n'roll era – and while his output over the years was variable, his energetic, genre-blurring early recordings will retain their appeal forever.

Ronald Wilson Reagan

1911–2004

American

Actor and 40th President of the USA

Ronald Reagan enjoyed the dubious distinction of being the oldest man ever elected US president. His genial manner and generally optimistic outlook made him popular with both voters and the right wing of his Republican Party. His negotiations with the Soviet Union in the late 1980s, resulting in the 1987 disarmament treaty did a great deal to reduce the tension of the Cold War.

Hailing originally from Tampico, Illinois, Reagan had a

Right: Reagan was a reasonably successful actor before he embarked on a career in politics. Here he appears as a deputy for his role in the 1953 remake of **Law and Order**.

Below: Gorbachev and Reagan sign a treaty eliminating US and Soviet intermediate- and shorter-range nuclear missiles, December 1987.

brief career as a sportscaster in Iowa before moving to Hollywood in 1937. He made his film debut the same year and appeared in some 50 films until his retirement from movies in 1964. Interested in politics from the late 1940s, when he was a Democrat, his involvement developed further during the 1950s when he was President of the Screen Actors Guild. In 1962 he officially became a Republican, endorsing Barry Goldwater in the 1964 presidential election against Lyndon B. Johnson. In 1966 he was nominated by the Republican Party as its candidate for the position of governor of California and was elected in the autumn of that year.

During his period as governor 1966–1974, he was strongly in favor of capital punishment and proved conservative in his reactions to student demonstrations and in social policy (with the exception of the decision — which he later regretted — to permit abortions).

In 1980, he won the Republican nomination for the presidency and, in securing victory against Jimmy Carter, became the oldest man ever to be elected US president. His first term was marked by economic problems, as he sought to deal with the high rate of inflation and unemployment that he inherited. From the end of 1982, when unemployment peaked,

Above: Nancy Reagan, Frank Sinatra, and President Reagan speaking at a campaign rally in Los Angeles in November 1984. Reagan brought a touch of Hollywood glamour to the White House.

Right: President Reagan saluting during the Pledge Of Allegiance at a Veterans Day ceremony at Arlington National Cemetery, Veterans Day 1988.

"Reaganomics," as his monetarist policies were known, brought sustained growth and a significant increase in employment. This, combined with the successful military intervention in Grenada in October 1983 (which had had the effect of helping restore US self-confidence after the trauma of the Iranian hostage crisis), helped Reagan to a virtual landslide in the 1984 presidential election.

As a conservative, Reagan — nicknamed the "Great Communicator" — was initially hawkish in his relations with the Soviet Union and the Warsaw Pact. The early years of his presidency saw a significant increase in Cold War tension, with new missile systems installed, despite considerable domestic opposition, in Europe and the concept of the Strategic Defense Initiative (or "Star Wars" as it became known) was developed to counter the perceived threat from the east. However, in Mikhail Gorbachev, the Soviet leader from 1985, he found a politician with whom he could negotiate, and over the next three years, Gorbachev and Reagan held four summit meetings

that resulted in a considerable reduction of mutual suspicion and the first meaningful moves to multi-lateral disarmament.

Having survived an assassination attempt in 1981, Reagan left office in 1989 having served two full terms. Although the Iran-Contra Scandal marred his second term, his reputation survived the affair and he left as one of the most popular presidents of recent times. Diagnosed with Alzheimer's disease in 1994, he died a decade later and was accorded a full state funeral.

Jackie Robinson

1919–1972

American

Baseball player and pioneering
African-American

An outstanding athlete, Jackie Robinson was the first African-American to surmount the color bar of major league baseball, when he signed for the Brooklyn Dodgers in 1947.

The youngest of five children, Jackie Robinson was born in Cairo, Georgia, moving with his family to Pasadena, California, the following year. Throughout his education he demonstrated considerable sporting prowess, and after leaving the University of California, Los Angeles in1941 he became a semi-pro footballer for the Honolulu Bears in Hawaii. Returning to the mainland, he intended to pursue a career in football with the Los Angeles Bulldogs, but the Japanese attack on Pearl Harbor in December 1941 and Robinson's subsequent draft into the army meant that these plans came to nothing. It

Above: Jackie Robinson was the first black player to join a major league baseball team.

Left: Robinson signing to play for the Brooklyn Dodgers. His jersey number was later permanently retired throughout major league baseball.

was in the army that he faced his first serious conflict with institutionalized segregation and racism.

His baseball career began in 1945 with the Kansas City Monarchs of the predominantly African-American Negro league. He attracted the attention of Branch Rickey, the general manager of the Brooklyn Dodgers, who was keen to find an African-American player capable of facing the undoubted racism that a non-white player would face in major league baseball without reacting angrily. A three-hour discussion was held in August 1945 between Robinson and Rickey, after which Robinson signed a contract and was publicly unveiled in October 1945 as a player for the Dodgers' linked team in the International League, Montreal Royals. He played the 1946 season for the Royals before making the move to the Dodgers the following year.

His debut for the Brooklyn team on April 15, 1947 was a seminal day for baseball. Robinson was the first non-white to break the sport's color bar since 1887, and his appearance was to cause considerable controversy. Receiving abuse from fans and opposition players, Robinson could not even rely on the support of his own teammates. It was only the backing of the Dodgers' management and of the National League President and other senior officials — who threatened to suspend any team that refused to play the Dodgers if Robinson played — that secured his position during this opening season. Such was Robinson's success on the field that he was to receive the Rookie of the Year award for 1947.

Having started his baseball career at the relatively old age of 28, Robinson's playing career was short-lived, and he retired from the sport in 1956. Having paved the way for non-whites to play baseball in 1947, subsequent seasons were to prove less controversial as the number of African-American players increased regularly from 1948 onwards. During his playing career he achieved a number of notable awards and, since his death in 1972, he has received a number of posthumous honors and other marks of recognition. Since 2004, April 15 has been known within baseball as Jackie Robinson Day in recognition of his pioneering involvement in the sport.

Above: A video tribute to Robinson at Shea Stadium in 2008. Players and coaches of both teams wore the number 42 in Robinson's memory.

Franklin Delano Roosevelt

1882–1945

American

32nd President of the USA

The only man to be elected president four times, Roosevelt led America through one of the most challenging periods of the 20th century, helping the country to recover from the economic depression of the 1930s, and steering the USA through the Second World War.

Born in Hyde Park, New York State, Roosevelt came from a well-off family whose antecedents were Dutch on his father's side and French on his mother's. Educated at Harvard, he met his future wife Eleanor, a distant cousin, whilst still a student before marrying her in 1905.

Called to the New York Bar in 1907, Roosevelt served as a New York state senator between 1910 and 1913 before becoming an Assistant Secretary to the Navy until 1920. In 1920 he was the Democratic nominee for the vice-presidency. However, in 1921, he was struck down by polio, which left him paralyzed from the waist down and he initially withdrew from public life.

Encouraged by his wife to return to public life despite his disability, Roosevelt was elected the 44th governor of New York in 1928, and in 1932 was selected as Democratic candidate for the presidency. Defeating Herbert Hoover, he became the 32nd President of the USA. He exuded goodwill and confidence, assuring the nation in his inaugural address that, "the only thing we have to fear is fear itself."

Faced by undoubtedly the worst economic circumstances in

Above: Roosevelt was born into a wealthy patrician New York family and married Eleanor Roosevelt, a distant cousin.

the nation's history, Roosevelt launched the New Deal, a wide-ranging program designed to provide relief for the unemployed and the elderly (through the Social Security Act of 1935), the rejection of the gold standard, and the devaluation of the dollar along with reform of the banking and economic systems. This he achieved through the creation of a number of agencies, some of which continue to play a pivotal role in the US economy 70 years on. Through a program of public works, such as Hoover Dam (completed in 1936), unemployment was reduced and the nation's infrastructure improved immeasurably.

Such was the success of the New Deal that Roosevelt, often known as FDR, was re-elected in both 1936 by a landslide, and in 1940, when he faced his second great challenge, World War II. Although Roosevelt's inclination was to keep the USA out of the war, his undoubted support for the Allied cause after September 1939 and his close relationship with Winston Churchill, who was to become British prime minister in 1940 resulted in much needed military equipment being shipped to Britain prior to the official US entry into the war in December 1941, despite opposition at home. In August 1941 he and Churchill issued the Atlantic Charter, committing their countries to "the final destruction of Nazi tyranny." The USA became the "arsenal of democracy", the most important financier and arms supplier of the Allies.

Re-elected for a fourth term in 1944, Roosevelt co-operated closely with Churchill and the Russian leader Stalin to construct a political framework for the postwar world, but he died in April 1945, shortly before the war's end. He was mourned around the world, but his legacy as one of America's greatest presidents lives on.

Above: Casablanca, Morocco, 1942: left to right, General Henri Giraud, President Roosevelt, General Charles de Gaulle, and Sir Winston Churchill.

Below: Roosevelt with King Abdukl Aziz of Saudi Arabia, chief of staff Leahy (kneeling), and Col. Eddy aboard a U.S. warship, February 20, 1945.

(Anna) Eleanor Roosevelt

1884–1962

American

First lady, social reformer and diplomat

The niece of Theodore Roosevelt, the 26th President of the USA, Eleanor Roosevelt was born in New York City to a family of considerable wealth, although both her parents died by the time she was 10 and she was brought up thereafter by her maternal grandmother. In 1889 she was sent to school at Allenswood Academy in England, where she became influenced by feminist and independent thinking. She married her distant cousin Franklin in 1905 and they had six children. Although the marriage had been a happy one, it came close to collapse in 1918 when Eleanor discovered that her husband had been having an affair with Lucy Mercer, her social secretary. Following his debilitating illness of 1921, she actively encouraged Franklin to return to public life and supported him throughout the rest of his political career.

It was during the 1920s that she first became actively involved with social affairs, work which was to dominate much of her life thereafter and which was used often to support her husband's political and economic ambitions. During the 1920s she worked with the Women's Trade Union League supporting its campaigns against child labor, a reduced working week and a minimum wage. She also fostered much closer links with the women in the Democratic Party, which undoubtedly aided Franklin's electoral success.

Historically, the role of the First Lady was largely ceremonial; Eleanor was, however, to bring considerable energy to the position and court controversy at the same time. In particular, she was prominent in her support of the civil rights movement, despite potential opposition from the Democrats in the southern states.

Her activity increased immeasurably after the US entered World War II in 1941. In 1941 she became Assistant Director of the Office of Civilian Defense and traveled widely, both with her husband and on her own, to wartime conferences and to bolster support from non-aligned countries. She was also prominent again in the furtherance of the rights of women and African-Americans during the war, most notably actively supporting the Tuskegee Airmen in their attempt to become the first black combat pilots.

Arguably, however, her greatest contribution was to come after her husband's death in her role as US representative to the newly formed United Nations between 1946 and 1952. Invited to take the position by President Harry S. Truman, she was to be a highly influential figure during these formative years. She was the first chairman of the UN Human Rights Commission and was one of those who drafted the UN's

Left: Eleanor Roosevelt wrote books, involved herself in feminist causes, and founded the Liberal group Americans for Democratic Action.

Above: For her political work Eleanor Roosevelt is immortalized in statue at the Franklin Delano Memorial in Washington, DC.

Right: Roosevelt listening intently to a speech via earphone, at the United Nations, New York, in September 1961.

Universal Declaration of Human Rights, which was adopted by the General Assembly in December 1948.

Leaving her post at the UN in 1953 following the election of President Eisenhower, she continued to be active in liberal politics thereafter, supporting the New Deal Coalition and women's rights, although she opposed the Equal Rights Amendment in a belief that it would not positively improve the position of women in US society. Her last public role, held between 1961 and her death the following year, was as chairman of the Presidential Commission on the Status of Women, a role to which she had been invited by Present John. F. Kennedy.

Babe Ruth

1895–1948

American

Baseball player

George Herman Ruth, universally known as "Babe" Ruth was one of the greatest and earliest sporting superstars, and still remains an American baseball legend whose playing statistics some eight decades after his heyday, sit at the upper reaches of the game. For the 100th anniversary of professional baseball in 1969, a popular ballot named Babe Ruth as the greatest player ever.

In the 1920s Babe Ruth was watched by thousands, and with his powerful swing he is credited with turning baseball from a speed-dominated, low scoring game, to a thrilling, high-scoring power game. During his 22-year professional career, Ruth hit 714 home runs with a batting average of .342, played for three teams and was seven times World Series champion. He was a big, powerful, left-handed player who started as a pitcher, but changed mid-career to become a phenomenal batter. Ruth's popularity in Japan helped the establishment of the Japanese professional baseball league. But he had two principle flaws, a tendency to over-eat, especially hot dogs and soda pop, and an inflammatory

temper which earned him suspension on more than one occasion.

Ruth's pitching potential was spotted during a school game in 1913, and aged 19, he was signed to the minor league Baltimore Orioles for $250 a month by the owner Jack Dunn. The other players called him "Jack's newest babe", soon shortened to Babe Ruth. The following year Ruth was sold to the Boston Red Sox but he made little impact until the 1916 season, when he made nine shutouts, an American League record for left-handers that stood until it was equaled in 1978.

In 1919 Ruth was sold to the New York Yankees after demanding that his $10,000 salary be doubled and refusing to play until it was. He was traded for a total of $100,000, then stayed with the Yankees for 15 years. In 1920 he hit 54 home runs at an average of .847, a major league record until 2001.

In 1921 Ruth achieved a career high, with a total of 59 home runs: he had become the biggest draw in New York and so many came to watch that the Yankees could afford to build the might Yankee Stadium, which became known as the "House that Ruth Built." Later that season the Yankees won their first World Series title. The Yankees won again in the 1928 World Series against the Cardinals when Ruth batted the second highest average in World Series history of .625. By 1930 he was earning a salary of $80,000, $5,000 more than US President Hoover.

Above: Babe Ruth demonstrating his classic swing in Yankee Stadium, also known as the "House that Ruth Built," (because he was such a huge attraction) before a game in 1921.

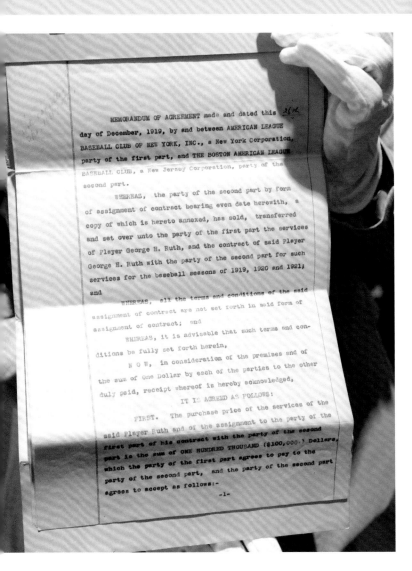

Above: The history-making contract which sold Babe Ruth from the Boston Red Sox to the New York Yankees in 1919.

Above right: Baseball signed by Babe Ruth. In 1969, a popular ballot named Babe Ruth as the greatest player ever.

Right: The bat with which Ruth hit the first home run in Yankee Stadium. The Yankees bought him in 1919 for $100,000.

In his declining years in February 1935, Ruth was traded to the Boston Braves as player and assistant manager. He played his last game on 30 May 1935 in Philadelphia and announced his retirement two days later. In 1946 he was diagnosed with throat cancer, and although initially chemotherapy helped, he died on 16 August 1948, aged 53. His body was laid out at Yankee Stadium and over 100,000 people came to pay their respects.

Ernest Rutherford

1871–1937

New Zealand

Physicist

Ernest Rutherford was a remarkable physicist and chemist whose work in diverse fields makes him one of the greatest research scientists. Rutherford's most important discovery was to determine the structure of the atom, and in the process became the first person to successfully convert nitrogen into oxygen by splitting the atom. Through his extraordinary experiments, Rutherford explained the attributes and complexities of radioactivity as the spontaneous disintegration of atoms: disproving the previously accepted Ancient Greek concept of atoms existing as stable entities.

Rutherford was the first scientist to detect individual nuclear particles (radioactivity) by electrical means, which led to the Rutherford-Geiger detector. Additionally, he laid the path for the modern smoke detector in 1899 while at McGill University when he blew smoke into his ionization chamber and observed the change in ionization: with just this observation he has helped to save the lives of thousands who would otherwise have died in house fires.

Rutherford was the child of a Scottish father and English mother who met after emigrating to New Zealand in the mid-19th century. He was born in rural Spring Grove, NZ in August 1871 and went on to attend Canterbury College, University of New Zealand, culminating in 1893 with a double first class honors in mathematics, mathematical physics and physical science. Yet he was unable to get a job as a schoolteacher and had to return to higher education at Canterbury College.

Rutherford had already made his name at the forefront of electrical technology before he traveled to Cambridge University, England, to take up a scholarship postgraduate position in 1895. He was soon devoting all his research to radioactivity and within two years had isolated two distinct types of emissions from radioactive atoms (thorium and uranium) that he called alpha and beta rays.

Right: Rutherford when he was President of the British Association in 1925. His work in splitting the atom changed the face of physics.

In 1898 Rutherford moved to McGill University in Montreal, Canada where he became professor of physics and discovered the radioactive gas radon. Working with chemist Frederick Soddy he investigated the recent discovery of radioactivity, which led them to propose that radioactivity results from the disintegration of atoms. This led to his Nobel Prize for Chemistry in 1921.

In 1907 Rutherford returned to England to take up the post of professor of physics at Manchester University. During World War I he stopped his research project and turned his attention to submarine warfare, in particular the study of underwater acoustics and was unable to return to physics until after the war. He then embarked on a long series of experiments during which he discovered that the nuclei of certain light elements (such as nitrogen) could be disintegrated by the impact of energetic alpha particles originating from a radioactive source. During the process fast protons are emitted: to do this he created the first artificially induced nuclear reaction and introduced nuclear physics to the world.

In 1919 Rutherford became professor of experimental physics and director of the Cavendish Laboratory at Cambridge. He was an inspirational and generous teacher and many of his students went on to become leading scientists in the next wave of physicists. Ernest Rutherford was knighted in 1931. He died on October 19, 1937 and was buried in Westminster Abbey.

William Shakespeare

1564–1616

English/British

Influential playwright, dramatist, poet and actor

Above: The plays of William Shakespeare have been regularly performed ever since he wrote them in the late 16th century.

Left: The First Folio edition of Shakespeare's' plays, printed in 1623, has been described as the most important book in English literature.

William Shakespeare is universally regarded as the greatest dramatist in the English language and his plays have been performed around the world for over 400 years. The density of his prose and historical analogies can be mystifying to his general audience, but his universal themes of love, hope, death, of battles for power and flights of pure fantasy, transcend the difficulty of the language. His continuing popularity endures through the beauty of his words and the liveliness of his work. The stories and the structures of his plays can be transplanted to any time and any place and completely reinterpreted, and yet they still transmit their magic. Shakespeare left a legacy of 38 plays, 154 sonnets, and a number of poems including two long narratives.

Although he was a prolific writer William Shakespeare did not record any details about his personal life. Accordingly, most of what we know about him comes from court records and other official documents. Shakespeare was born in April 1564, possibly on the 23rd, in Stratford-upon-Avon and was the third child and first son of John and Mary Shakespeare. As his work shows, he was clearly well educated with a good knowledge of the classics, probably at the King's New School in Stratford. He married Anne Hathaway when he was 18 and she 26, and had three children. Shakespeare moved to

London in the late 1580s to further his career as an actor and playwright, although he came home to Stratford as often as practicable.

The order in which Shakespeare wrote his plays is unknown, but he was popular and successful with both the ordinary theatergoer and the royal court. Between 1592 and 1594 plague closed the London theaters, so Shakespeare wrote poetry instead. 154 sonnets dealing with love, lust, separation and despair, betrayal and death have survived: one group is addressed to a young man and another to a "dark lady," but scholars have not yet identified anyone by name, and indeed, they may not refer to specific people at all.

In 1594 the theatres reopened and in the following four years Shakespeare wrote five history plays, six comedies and, a tragedy, *Romeo and Juliet,* all of them wildly successful. In

Above: The Globe Theatre was reconstructed in the late 20th century near its original site beside the River Thames in London.

Below: Shakespeare's work has often been filmed. Here Michelle Pfeiffer stars in the movie William Shakespeare's A **Midsummer Night's Dream**."

1599 the Globe theatre was built on Bankside, outside the confines and restrictions of the City of London, and its huge stage allowed the performance of more spectacular plays. Shakespeare was part owner of the company and the proceeds made him a wealthy man. The Globe opened with *Henry V*, and performances of *Julius Caesar*, *Twelfth Night*, *Hamlet* and *Othello* followed. James I became king in 1603 and increased the royal patronage of the theater; tactfully, the company changed its name to become the King's Men.

By 1612 Shakespeare returned home permanently to Stratford where he now had grandchildren. He died at home on 23 April 1616 and was buried in the Holy Trinity Church, Stratford.

Shakespeare's genius lay not only in his amazing intellectual capacity, but also in his imagination, poetic power and ability to create vivid characters and situations, ensuring that his work remains, in the words of his contemporary Ben Jonson "not of an age, but for all time."

Steven Spielberg

Born 1946

American

Film director

Above: In 1994 **Schindler's List** won two Academy Awards for best director and best picture.

egarded as one of the most influential figures in the movie industry and listed as one of the 100 most important figures of the 20th century by *Time* magazine, Steven Spielberg is one of the most successful movie directors of all time.

Born in Cincinnati, Ohio, Spielberg's father was an electrical engineer and his mother a restaurateur and concert pianist; following his parents' divorce, he moved with his father to California. Interested in film from an early age — he produced, for example, a nine-minute film to earn his photography merit badge in the Boy Scouts — he sought to study this at college but failed to gain admission to the University of South California's School of Film, Theater and Television before being accepted to read English at California State University.

Initially Spielberg worked on television projects, with his first directorial credit being part of the pilot for *Night Gallery* in 1969. This led to further TV work including three TV movies and his first movie, the unsuccessful *The Sugarland Express*. His big breakthrough came when he was selected to direct the movie of Peter Benchley's novel *Jaws*. Although almost canned during production, the film's release saw it set a record for takings at the US box office and shot Spielberg to the top of his profession. The film was nominated for Oscar as Best Picture, the first of many nominations that Spielberg was to achieve over the years. It was not, however, until *Schindler's List* in 1994 that Spielberg finally achieved a success in the Academy awards for Best Director. The film also won the Oscar for Best Film, and Spielberg repeated his Best Director award with *Saving Private Ryan* four years later.

The success that *Jaws* brought Spielberg was to allow him a certain latitude in the future, and the subsequent years were to result in some of the greatest films — in several genres — ever produced. These included *Close Encounters of the Third Kind* (1977), which he also wrote; the quartet of films featuring Harrison Ford as Indiana Jones — *Raiders of the Lost Ark, Indiana Jones and the Temple of Doom, Indiana Jones and the Last Crusade* and, most recently, *Indiana Jones and the Kingdom of the Crystal Skull*; *E. T. The Extra-Terrestrial* (1982); *The Color Purple* (1985); *Empire of the Sun* (1987); *Jurassic Park* (1993); *Schindler's List* (1993); *Saving Private Ryan* (1998); *A. I. Artificial Intelligence* (2001); and *Munich* (2006). His movies

Above: Spielberg going over a scene with stars Anthony Hopkins and Morgan Freeman on the set of **Amistad.**

Right: Steven Spielberg and producer Kathleen Kennedy at the 20th anniversary premiere of blockbuster movie, **E.T. The Extra-Terrestrial.**

share certain characteristics, including inventive special effects, impressive musical soundtracks, and most of all, a gripping storyline.

Of these, perhaps the most significant was *Schindler's List.* Not only did it win Spielberg his first Best Director Oscar, but also reflecting his own Jewish roots, he used the profits to fund a non-profit organization, the Shoah Foundation, that records the often harrowing testimony of Holocaust survivors.

Aside from his role as one of the cinema's most commercially successful directors, Spielberg was also one of the three co-founders of the Dreamworks studio in 1994, along with Jeffrey Katzenberg and David Geffen. Dreamworks has produced many of Spielberg's own films, whilst also encouraging a range of both real-life and animated films from a number of other prominent directors.

Josef Stalin

1879–1953

Russian

Dictator and leader of the Soviet Union

Josef Stalin a totalitarian dictator, perhaps the most feared of modern times. He transformed Russia into a major world power during and after World War II and his influence on modern history was profound. However, his methods were brutal and callous, and under his regime millions of Russians died as a direct result of his policies.

Josef Vissarionovich Dzhugashvili was born in Gori, Georgia, and his mother wanted him to study for the priesthood. However he was expelled from the seminary at Tiflis and later

joined the Bolsheviks in 1903. He was arrested several times and was eventually exiled to Siberia where he languished until the Revolution in 1917. He was appointed Commissar of the Nationalities and was in charge of the defense of the important city of Tsaritsyn during the civil war that followed the Revolution. The city was later renamed Stalingrad in his honor.

He became Chairman of the Politburo after the death of Lenin in 1924, and he quickly squashed all opposition (real or imaginary) by arresting, imprisoning and often executing those who dared to oppose him. Despite the reign of terror, he also instituted a series of five-year plans that organized Russia's industrial potential and began the collective approach to agriculture, although the policies employed resulted in a severe famine.

In 1938 he extended the Great Purge of the Communist Party to the Red Army's officer corps, executing, demoting or exiling some 10,000 officers of all ranks. This was to cause serious problems when Russia became embroiled in World War II, although Stalin had already disconcerted the Allied powers by signing a non-aggression pact with Hitler in August 1939. Stalin was taken completely by surprise when Hitler attacked Russia in June 1941, and the virtually leaderless Red Army was no match for the German Panzers. Despite panic at the Kremlin, Stalin stayed in command of the government in Moscow, even as German troops entered the outskirts of the city. By this time he was commander of all Soviet

Below: Josef Stalin, President Truman, and Winston Churchill – the principal Allied leaders of World War II – meet at the Potsdam Conference in July 1945.

armed forces as well as being Chairman of the Council of People's Commissars. Fortunately there were still some able officers and Marshall Zhukov eventually led a successful counterattack in December 1941. Subsequently, a reorganized industrial base began turning out tanks, guns and aircraft in ever increasing numbers, while the seemingly inexhaustible supply of recruits eventually made the Red Army the largest and most powerful in the world. Nevertheless, the defeat of Germany involved some 20 million casualties, more than the casualty list for all the other combatant nations combined.

Russia's military power base was almost totally negated in 1945 by the arrival of the atomic bomb and Stalin set the highest priority on making the Soviet Union a nuclear power, which was achieved in 1949. In the years following World War II, Stalin extended his Communist rule to the countries of Eastern Europe, which the Red Army had liberated in the closing stages of the war. In Churchill's words, an Iron Curtain had descended across Europe, and it remained in place for almost half a century. Stalin himself became increasingly paranoid and eventually died of a stroke in March 1953, although some sources suggest he was poisoned by other members of the Politburo who feared for their own lives.

Below: The Soviet government in 1938: Stalin (center) and to his immediate right, Vyacheslav M. Molotov, his notorious foreign minister.

Above: Josef Vissarionovich Dzhugashvili, was born in Georgia and known to the world as Stalin, and sometimes ironically as "Uncle Joe."

Marie Stopes

1880–1958

British

Pioneer of birth control and family planning

Above: Through her work Marie Stopes made a considerable contribution to breaking down the ignorance and taboos surrounding sex and contraception.

Left: Stopes founded the first birth control clinic in Britain, then later focused on the needs of the developing world.

Maries Stopes was a campaigner and advocate of family planning and birth control, who established the first family planning clinics in Britain. Her work and writings did much to break down the taboos surrounding sex and contraception.

Born in Edinburgh, Stopes initially trained as a palaeobotanist, gaining degrees from University College, London and the University of Munich. In 1904 she became the first female lecturer in science to be appointed to the staff of Manchester University, before spending 18 months in 1907–1908 on a scientific mission to Japan exploring fossil plans and lecturing at Tokyo University.

Married in 1911 to Reginald Ruggles Gates, it was the failure of this union — which Stopes claimed had been unconsummated — five years later that was to lead her into a completely new direction, that of eugenics and family planning. The failure, which she ascribed to ignorance about sexual matters and contraception, led her to write her first book on the subject, *Married Love*, which was published amid considerable controversy in 1918. The book was initially banned in the USA when it was first published, such was the nature of its content. Other books followed, such as *Wise Parenthood*, also published in 1918, *Contraception: Its Theory, History and Practice* (1923), *Radiant Motherhood* (1920), *Sex and the Young* (1926) and *Sex and Religion* (1929); in all she published some 70 books on the subject between 1918 and her death in 1958.

Her second marriage in 1918 to Humphrey Verdon Roe was more successful with a son being born in 1924. Along with her husband she opened Britain's first family planning clinic at 61 Marlborough Road, Holloway, London in March 1921. This

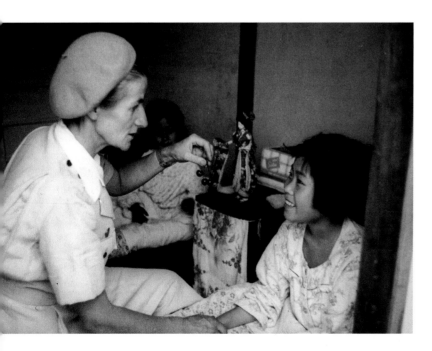

clinic offered free treatment to married women at a time when, prior to the establishment of the National Health Service, medical treatment was normally paid for. The Mothers' Clinic, as the establishment was known, moved to central London in 1925. This was the first of a number of clinics that were established; although the original business was forced into receivership in 1975, it was reformed the following year and today the Marie Stopes International organization has some more than 450 clinics worldwide.

Apart from offering pioneering treatment, the clinic was also at the forefront of research into contraception. It was this work, along with that of other pioneers, that helped to break down the lack of knowledge in sexual matters that had existed earlier.

Controversy was never far from Stopes during the latter part of her career and this was particularly true of her belief in eugenics, which was fashionable in intellectual circles at the time. In her 1920 book, Stopes propounded the view that certain types of people — such as the disabled — were genetically ill suited to procreation and ought to be sterilized. Such was her belief in this theory that she disinherited her own son when he married the partially sighted daughter of the scientist Barnes Wallis. Following her death, much of her estate was passed to the Eugenics Society.

Below: Dr Marie Stopes with her husband Humphrey Verdon Roe outside the Royal Academy in London in 1926.

Louis Sullivan

1856–1924
American
Architect

Known as the 'father of modernism," Louis Sullivan was a brilliant architect who spans the gap between Belle Epoch and Modernism and was responsible for nurturing the early talent of the young Frank Lloyd Wright for six years. He became the leading Chicago architect of his era and senior

Above: The Wainwright Building, at 709 Chestnut Street, St. Louis, Missouri. Built in 1890-91 and designed by Dankmar Adler and Louis Sullivan, it was one of the first skyscrapers in the world.

Left: A portrait photograph of Louis Sullivan, probably taken in the latter part of the 19th century.

Opposite: The Merchants' National Bank at 833 Fourth Street, Grinnell, Iowa, designed by Louis Sullivan, was declared a National Historic Landmark in 1976.

member of the Chicago School of Architects. One of the pioneering designers and developers of skyscraper construction and technology, he literally changed the way modern cities look.

Sullivan was the first architect to articulate what became the Modernist mantra "form (ever) follows function," meaning that if a building were properly designed for its purpose, it would, by extension, be beautiful. Accordingly, he did away with unnecessary embellishments, to create clean-lined buildings, although he did still like the selective use of organic decoration across the surfaces, particularly using terra cotta and cast iron. Such decoration he saw as the logical balance between nature and technology.

Skyscrapers were not technologically possible before the development in the second half of the 19th century of cheap mass-produced steel girders that were able to take the weight of the building off the walls. Skyscrapers also required the invention of the elevator to take people, goods and services to the higher reaches of buildings. Sullivan was the first architect to really consider the problems and solutions for tall, rather than long, buildings as he was working at the exciting epoch when technology for the first time made building upwards possible.

Louis Henri Sullivan was the son of an Irish immigrant father and a Swiss mother. He entered the Massachusetts Institute of Technology aged 16 to study architecture, but became bored after a year with classicism, so moved on to Philadelphia and a job with architect Frank Furness.

In 1871 the city of Chicago was almost destroyed by fire and the rebuilding offered exciting opportunities for young ambitious architects like Sullivan. He moved there in 1873 to a job with William Jenney, the first architect to work with steel-framed buildings. Before the year was out, however, Sullivan left to travel to Paris where he joined the École des Beaux-Arts for a year before returning to Chicago. In 1879 he joined civil engineer and architect Dankmar Adler's firm and within a year was made a partner. Their architectural practice soon established their reputation as modern office builders, with Louis Sullivan particularly specializing in devising a new type of building — the steel frame high rise.

The steel was constructed in "column frame," higher than ever before and with much bigger windows and therefore better light within the building. With the rebuilding of Chicago, Sullivan created a new grammar of form to create a high rise consisting of base, shaft and pediment. Sullivan's seminal

buildings include the Gage Building and the Stock Exchange in Chicago.

In the Panic of 1893 the construction business declined and Adler and Sullivan dissolved their partnership. It also marked the end of the Chicago School as a leading architecture style, and a return to traditional neoclassical forms. Sullivan rarely worked again. He slipped into alcoholism and died a lonely death in a Chicago hotel in 1924. Frank Lloyd Wright contributed to his funeral costs.

Margaret Thatcher

1925-2013

British

First female British Prime Minister

Margaret Thatcher earned her place in history as the first female prime minster of the United Kingdom from 1979 until 1990, winning three election victories, and, arguably, transforming Britain. A Conservative politician, she consistently advocated a reduction in state control, privatization of industry and tight control of public expenditure.

Born in Grantham, Lincolnshire, the daughter of Alderman Alfred Roberts, a grocer, Margaret Roberts progressed to Oxford University where she graduated in chemistry. She married Denis Thatcher, a successful businessman in 1951, and qualified as a barrister in 1954. Her strong political ambitions were realized in1959, when she was elected Member of Parliament for Finchley, in London.

Thatcher demonstrated efficiency and ambition in equal measure from the start of her parliamentary career and was appointed to several junior ministerial posts during the Conservative administrations of the 1960s. Following the Conservative election victory in 1970, she became Secretary of State for Education and Science. After two electoral defeats, the Conservative leader Edward Heath lost the faith of his party and Thatcher was elected leader in his place. Exploiting the unpopularity of James Callaghan's Labour government, Thatcher led the Conservatives to electoral victory in 1979.

Above: Margaret Thatcher celebrates her 1979 general election victory with husband Denis, outside the British Prime Ministers residence at 10 Downing Street, London.

Her first term as Prime Minister was characterized by a massive increase in unemployment as she sought to reduce the burgeoning rate of inflation, and by huge unpopularity. The military campaign to retake the Falkland Islands following their seizure by Argentina in April 1982 saved Thatcher's government. The victory, which was wrought at considerable cost, saw British national self-confidence restored and ensured her re-election in 1983.

Thatcher's second term was marked by more radical legislation, in particular the privatization of many state utilities, which created the concept of popular capitalism, with countless millions participating in the resulting share issues. However, the social revolution that Thatcher sought to engineer was not without its strains. In 1984 the Conservatives opposed the National Union of Mineworkers under the charismatic leader Arthur Scargill, and, in a year-long strike that polarized the country and ultimately broke the NUM, the government forcibly restrained the historically powerful trade union movement.

A further victory at the polls in 1987 saw Thatcher re-elected for a second time, but her popular touch, which had guided much of her success since the Falklands War, deserted her. A number of unpopular reforms — most notably the Community Charge (or 'Poll Tax' as it became known) — resulted in opposition both within her party and throughout the country at large. Although she survived to become the longest-serving prime minister of the 20th century, by 1990 her position became increasingly untenable and she was forced to resign.

In retirement, Thatcher's shadow continued to hang over the Conservative Party for a number of years. Although her

Above: Margaret Thatcher during a press conference in Jerusalem at the end of her visit in 1986, it was the first of a British prime minister to Israel.

Above right: Margaret Thatcher and Ronald Reagan waving to the press after their arrival in Camp David on 22 December, 1984.

Right: Margaret Thatcher addressing the press before handing her resignation as Prime Minister to Queen Elizabeth II on November 28, 1990.

domestic political role was diminished, Thatcher's influence abroad remained strong, particularly amongst the emergent democracies of central and Eastern Europe through her Thatcher Foundation. Interestingly, her resolve and strength of character remain widely admired among politicians of all parties, and in her twilight years she is respected as a prime minister of conviction who, at the very least, got things done.

Mother Teresa

1910–1997
Macedonian/Indian
Religious humanitarian and missionary

Mother Teresa felt her vocation to devote her life to the Roman Catholic Church at the age of 12, but it was only after teaching at a convent in Calcutta (now Kolkata) that she discovered her true calling — to help those who no one else would touch, the desperately poor, sick, and undervalued people of the slums. She devoted her life to her mission and even created her own religious order, the Missionaries of Charity, to take her work around the world.

Agnes Gonxha Bojaxhiu — later known as Mother Teresa — was born in Skopje, Macedonia on August 26, 1910. Of Albanian descent, she felt the call of the Church after being inspired by stories of missionaries and their works. At 18 she left home to join an Irish community of nuns known as the Sisters of Loreto and never saw her family again. In preparation for her calling she was sent to Loreto Abbey near Dublin where she also learned to speak English, the language of the mission.

Agnes arrived as a novitiate at her first posting at Darjeeling in 1929 and two years later she made her first vows and took the name Teresa, after St Teresa of Lisieux, the patron saint of missionaries. After moving to Calcutta she taught at St Mary's High School and there, on 14 May 1927, she took her final vows.

Outside the school lay the heaving slums full of illiterate, desperately poor and completely overlooked people. Their plight got even worse after the famine of 1943 and the sectarian violence during Indian Partition in 1946. Mother Teresa increasingly felt that this was her true calling, and in 1948 she at last received permission to leave the school and devote herself to the people of the slums. She also took Indian citizenship.

Although lacking any funding, Mother Teresa set up an open-air school in Motijhil for the children of the slums, but she had to beg for supplies and food. However, as her work became known, she was joined by volunteer helpers and aided by financial contributions.

In 1950 Pope Pius XII granted Mother Teresa permission to start her own religious order with 13 members in Calcutta. In time this lead to the creation of the Missionaries of Charity. By the 1960s the order had orphanages, hospices and leper shelters across India, and within a decade Mother Teresa had

Left: Mother Teresa was beatified in 2003, only six years after her death. This is the last step before canonization as a saint, and took place more quickly than for any other individual.

become an international celebrity.

The order became an International Religious Family with active and contemplative branches around the world after Pope Paul VI issued the decree in 1965. In 1979 Mother Teresa won the Nobel Peace Prize for her work with the poor, orphaned, sick and dying.

In 1983 Mother Teresa suffered a heart attack and continued to suffer from heart troubles for the remainder of her life. She died of another heart attack on 5 September 1997.

By the time she died, the Missionaries of Charity were operating 610 missions in 123 countries. Mother Teresa was quickly beatified by Pope John Paul II and given the title Blessed Teresa of Calcutta.

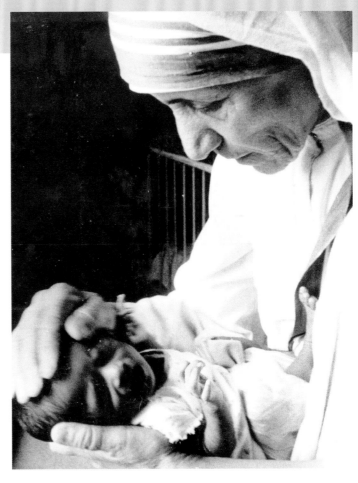

Right: Mother Teresa devoted her life to helping the sick and underprivileged of first Calcutta, then the rest of the world.

Below: Mothers and children at the Mission in Calcutta. The Missionaries of Charity help orphans, lepers, and the sick and dying.

Juan Trippe

1899–1981

American

Airline founder

Juan Trippe was an aviation pioneer and the founder of Pan American World Airways, one of the great airline companies of the middle years of the 20th century. He was an astute businessman who did much to make global travel affordable by the many not the few.

Trippe's studies at Yale were interrupted by the US entry into World War 1, in which he served as a night bomber

Below: Juan Trippe in London, 1946, soon after the launch of his controversial London–New York economy fares.

pilot for the US Navy. This wartime experience undoubtedly influenced his choice of career following his graduation from Yale in 1922, although initially he began work on Wall Street.

In 1923 he joined New York Airways, a company that provided an air-taxi service for the wealthy of New York. Shortly afterwards he and some friends invested in a small airline — Colonial Air Transport — which won the first US air mail contract between New York and Boston, before establishing the Aviation Company of the Americas, based in Florida, that was intended to develop business in the Caribbean. This company was to evolve into Pan American Airways — later better known as Pan Am — which was to fly its first flight on October 19, 1927 from Key West, Florida, to Havana, in Cuba.

Pan Am expanded during the succeeding years through a number of significant acquisitions, and the development of the airline as a global entity owed much to Trippe's belief in the democratization of flying — that it was something that all should be able to enjoy rather than just the well-healed elite. In the 1930s the company had pioneered long-distance flights over the Atlantic and Pacific oceans using its fleet of 'Flying Clipper' flying boats, but it was in the immediate postwar years that Pan-Am demonstrated this philosophy. Under Trippe's guidance, Pan Am launched the first economy fares in 1945, initially on the route from London to New York. Such was the opposition to the move, however, that British airports were barred to Pan Am flights with these tickets, and the airline was forced to use Shannon Airport in Ireland. During these years he was also instrumental in the development of Pan Am's non-aviation business interests, including the

Above: First Lady Mamie Eisenhower launching Pan Am's Boeing 707 **America** at National Airport in 1958 while Juan Trippe looks on.

Below: Juan Trippe, founder and President of Pan American Airways, in 1947. He was posthumously awarded the U.S. Medal of Freedom.

creation of the InterContinental chain of hotels.

Technically astute, Trippe also recognized the potential of the jet aircraft and Pan Am was amongst the first airlines to fly both the Boeing 707 and the Douglas DC-8 aircraft; the company's first jet flight took place in October 1958. The launch of the jet age allowed Trippe to increase further the number of passengers that he could carry at lower fares.

However, perhaps his most significant contribution to the airline industry was the concept of the Boeing 747, which he championed because he believed that there was scope for an aircraft with a greater capacity than the 707. Although he believed that the future lay in supersonic flight and saw the 747 primarily as a freight aircraft in the future, Trippe's involvement ensured that the 747 was produced, with Pan-Am was the model's launch customer.

Trippe retired as president of the airline in 1968. His contribution to world aviation was marked in 1965 by the receipt of the Tony Jannus Award and, in 1985, he was posthumously awarded the US Medal of Freedom.

Desmond Tutu

Born 1931

South African

Anglican religious leader and anti-apartheid campaigner

Archbishop Desmond Tutu is a South African cleric who was an outspoken opponent of the apartheid regime. His role in bringing peace to the country after the divisive years of apartheid has done much to smooth the country's transition to a "rainbow nation."

Tutu was born in Klerksdorp, a town to the southwest of Johannesburg, and was the son of a primary school headmaster. He went on to study theology at the University of South Africa and then completed his education at the University of London. He briefly followed in his father's footsteps working as a teacher but in 1960 became a parish priest. Promotions followed swiftly and he was made the bishop of Lesotho, a self-governing enclave within South Africa, in 1977. He held the position for the next two years, before returning to his homeland to take up the post of the secretary-general of the South African Council of Churches. In 1984 he was made the bishop of Johannesburg, becoming the first black African to hold the post in the process.

South Africa was a divided society, one split along racial lines. The system of apartheid (separateness) had been introduced in 1948, when the Afrikaner Nationalist Party

Above: In 1981 Desmond Tutu was a bishop in the Anglican Communion and General Secretary of the South African Council of Churches.

came to power. Racial segregation had existed before, but apartheid legalized and codified it. Black Africans were derived of political and civil rights, had poorer access to facilities of all types, and lived in separate townships, which were often little more than slums. Opposition to apartheid took many forms, both violent and peaceful, and Tutu emerged as one of its most vocal and persuasive opponents.

One of the controversial means of ending apartheid was the use of international sanctions. Critics believed they would mostly harm black South Africans; supporters argued that they were the quickest way to bring about change. Tutu was a very vocal supporter of their introduction, even though his stance meant that he frequently risked imprisonment by the authorities. He was also firmly opposed to the use of violence to bring change and argued that any political solution had to involve peaceful negotiation between black and white. His efforts to bring about peaceful change were recognized in 1984, when he was awarded the Nobel Prize for peace and two years later he became the first black archbishop of Cape Town.

Apartheid effectively came to an end when the reform-minded President F. W. de Klerk lifted the ban on the main black opposition movement, the African National Congress, and released several of its leaders from jail, including Nelson Mandela, in 1990. The various structures of apartheid were dismantled and a new multi-racial constitution thrashed out over the following months. Mandela became the country's fist black president the following year. Although

Above: The enthronement of Desmond Tutu as archbishop of Cape Town at St Georges Cathedral, in September 1986. He was the first black man to hold this post.

Right: Nelson Mandela receiving the five-volume Truth and Reconciliation Commission final report from Archbishop Desmond Tutu, in October 1998.

South Africa was by-and-large a functioning democracy, the wounds of apartheid ran deep and could not be healed easily. The decision was taken to create the Truth and Reconciliation Committee, a body that was to help both black and white come to terms with the past. Tutu, whose moral authority was unimpeachable, was its chairman between 1995 and 1998, thus confirming his position as one of the leading figures in modern South African history.

Lech Walesa

Born 1943

Polish

Trade union activist and politician

Lech Walesa was a charismatic Polish trades union leader, who formed Poland's first independent trade union in the face of Communist opposition. He went on to become President of Poland, 1990–1995.

Walesa was born into a country that had been liberated by the Red Army in 1945 but had immediately been turned into a one-party Communist state that was dominated by and, to

Right: After his landslide victory in 1990 President-elect Walesa took the oath of office in front of the Polish National Assembly.

Below: President Lech Walesa kisses the hand of Pope John Paul II, while holding the first copy of Poland's 200-year old constitution.

a considerable degree, controlled by its large neighbor to the east, the Soviet Union. Walesa grew up in the town of Popowo, the son of a previously well-to-do family that had been impoverished by the destruction wrought on Poland during World War II. He worked as an electrician in Gdansk's Lenin Shipyard between 1966 and 1976, became a trade union organizer, and led its strike committee in 1970. That year, a wave of industrial action, including a strike at the Gdansk shipyard, forced the resignation of the Polish head of state, Wladyslaw Gomulka. Walesa, who was becoming a well-known union leader, was sacked in 1976 for further strike activity and was dismissed on similar grounds from his next job three years later.

Edward Gierek, Gomulka's successor,

was forced to step aside in 1980, largely because his attempt to modernize the country's industrial base had plunged it into severe debt. On August 14 1980, Walesa became leader of a strike in the Lenin shipyard which was protesting about food prices, and which sparked more industrial action across Poland. In September, the Communist government was forced to sign an agreement with the Strike Committee allowing the formation of free trade unions. The Strike Committee was transformed into a national federation of unions with some 10 million members and became known as *Solidarnosc*, or Solidarity, with Walesa as chairman.

In December 1981 a tougher government headed by General Wojciech Jaruzelski, attempted to re-impose state control over Poland in 1981 by imposing martial law. Solidarity was banned and its leaders were held in detention or kept under surveillance. Walesa was detained until November 1982.

Jaruzelski suspended martial law towards the end of 1982 and it was lifted in mid-1983, but he had, in fact, overplayed his hand by his draconian measures. The unrest continued, and dissidents were given international support, not least by the Polish Pope John Paul II, who visited his homeland that year and was received by huge crowds. Walesa was granted a private audience with the pope and was also awarded the Nobel Prize for Peace in October 1983.

Matters came to a head in 1989,

Above: Lech Walesa worked as an electrician in the Lenin shipyard where he founded the Solidarnosc (Solidarity) Trade Union in 1980.

when, with the collapse of Communism throughout Eastern Europe, the ban on Solidarity was lifted. Walesa was now able to negotiate with Jaruzelski over the future path of Poland. Agreements were reached on reforming the country's political system, not least in allowing for free and fair elections. Walesa was elected president in a landslide victory in 1990. He guided Poland to a free-market economy system, but his period in office was dogged by controversy and political infighting and he stepped down in 1995. Nevertheless, his role in brining about the democratization of his homeland was pivotal.

Andy Warhol

1928–1987

American

Artist and filmmaker

One of the prime movers behind the Pop Art movement of the 1960s, Andy Warhol was an innovative and creative artist whose mass-produced art and involvement in avant-garde circles made him famous. Noted saying that "In the future, everyone will be famous for 15 minutes," his celebrity has lasted considerably longer.

Warhol was born Andrew Warhola to parents of Czech origin in the city of Pittsburgh, Pennsylvania, and studied at the Carnegie Institute of Technology. By the late 1950s there was a change in the direction of the contemporary art scene in the United States with a movement from Abstract Impressionist to a new form known as Pop Art. This focused on the depiction of everyday images drawn from advertising, mass media, famous personalities and pop culture. It was christened Pop Art as a nod to the growing domination of popular culture that was manifest in the years after World War II. Warhol did not found the movement – that is usually ascribed to Jasper Johns and Robert Rauschenberg – but he became its greatest exponent, not least for his mass production approach to the form and his abilities as a self-publicist.

Warhol's breakthrough as the leading Pop Artist of his generation came in 1962 when he produced a series of silkscreen prints of mundane, everyday items, such as cans of Campbell's soup and Coca-Cola bottles, or leading film stars The movie stars were motifs that he would return to again and again, with figures such as Marilyn Monroe and Elizabeth Taylor. He worked from his New York studio, the appropriately named Factory, which became a magnet for the rich and famous, as well as the various hangers-on that the artist attracted. Warhol was also closely involved with one of the leading experimental bands of the day, the Velvet Underground, and designed the groundbreaking Pop Art cover for their first album, *The Velvet Underground and Nico* in 1967.

Warhol was also an avant-garde underground filmmaker, and his earlier experiments in the genre were centered on attempts to remove the individuality of the director from the process of actually creating the movie. Warhol achieved this by using a camera fixed in one position on an object and filming without sound. *Sleep* (1963), a three-hour movie, best exemplifies this technique. The artist later began to use sound, but dispensed with the fixed camera to produce movies with long single takes but ones that employ frequent and violent changes of the viewer's visual and aural perspectives. *Chelsea Girls* (1966) is the best example of this technique. Warhol was

Left: Warhol signing copies of his book **From A to B and back again** in London, in November 1975.

Above: **Turquoise Marilyn** *is made from acrylic and silkscreen on linen, and is one of Warhol's most iconic works.*

Right: **Campbell's Soup Cans** *(1962) is one of Warhol's most famous images.*

shot and wounded by one of his movie performers in 1968 and thereafter, as he became more withdrawn, he gave away considerable control of his output to Paul Morrissey, an ex-assistant and cameraman. With Morrissey's input the movies became more mainstream and exploitative with titles such as *Flesh* (1968), *Trash* (1969), *Frankenstein* (1973) and *Dracula* (1974).

Warhol became increasingly reclusive in his latter years but continued to be highly productive. He died on February 23, 1987 aged 56 following complications after routine gall bladder surgery in New York. Since his death his works have continued to draw high bids at auction and interest in him remains high.

David Warren

1925-2010
Australian
Scientist

David Warren is one of the unsung heroes of the world of aviation, the man who invented the "black box" recorder, which itemizes the instrument readings and conditions of an aircraft in flight. This data is vital in unraveling the causes of an air crash.

Although born on a remote mission station in Northern Australia, David Warren was educated in Sydney. His life was torn apart in 1934, when his father was killed in an air disaster in Southern Australia. The desire to know what happened during those final moments helped shape his career. One of the last presents his father gave him was a crystal radio set, which he listened to in the school dormitory, and which spawned his interest in electronics.

In 1944 he became a science teacher – a stepping-stone to the post of lecturer in Chemistry at the University of Sydney. However, his real interest lay in research rather than teaching, and in 1948 he became a Scientific Officer at the Woomera Rocket Range. Five years later, after earning his doctorate, Warren accepted the post of Principal Research Scientist at the Aeronautical Research Laboratory in Melbourne. This placed him in the heart of Australia's scientific establishment.

In 1953, Warren was called in to investigate the crash of the Comet, the world's first jet-powered commercial airliner. The cause of the accident remained a mystery, so Warren argued

Left: David Warren with an early version of his life-saving device.

Below: David Warren's invention, the flight data recorder, enables crash investigators to discover the cause of air accidents.

that some form of cockpit-mounted electronic voice recorder could help explain the cause of future accidents. When his suggestion was ignored, he set about building his own voice recorder – the prototype of "the black box". His model was a data recorder he saw at a trade fair, but Warren came up with his own design, tailored to the specific needs of aviation.

His first design could store up to four hours of voice recording, together with a matching set of instrument readings. It was designed to wipe out older readings as it went along, so it always recorded the last few hours of an aircraft's flight. The recorder was stored in a crash-proof and fire-proof box, which could be retrieved in the event of an accident. In its first form it wasn't black at all, but was known as the "Red Egg."

It took five years for Warren to convince his government that the device was of value, and it was first tested during a flight to London in 1958. It was another five years before it

Above: Black box instruments, including the cockpit voice recorder, recovered from the wreckage of a Libyan Arab Airlines Boeing 727, in 1973. Flight data recorders are usually mounted in the tail section, where they are most likely to survive a crash.

became a compulsory part of Australian civil airline equipment, its adoption spurred by another Australian air crash in 1960. By this time the aviation industry had come to recognize the importance of Warren's invention, and it rapidly became an essential tool for the world's airlines. Today a modernized version of Dr. Warren's "black box" is used in all commercial passenger aircraft. While the "black box" or aircraft data recorder can never directly save an aircraft from disaster, it created a tool that allowed aircraft designers and operators to learn from their mistakes. As a result, Warren's black box flight data recorder made flying much safer, and helped pave the way to a new era of air travel.

George Washington

1732–1799

American

Soldier and first president of the USA

George Washington's legacy and lasting influence upon America is all the more remarkable because his life involved two very different but supremely important roles – those of military commander and pioneering statesman. He was the first president of the United States of America, a highly respected statesman and soldier regarded as the "father of his country."

Washington was born a loyal son of the British empire in Westmoreland County, Virginia. At the age of 16 he was appointed surveyor of the estates of Thomas, Lord Fairfax, and acquired valuable knowledge of the state, which he would

Above: The first president of the United States was both a successful military commander and a pioneering statesman.

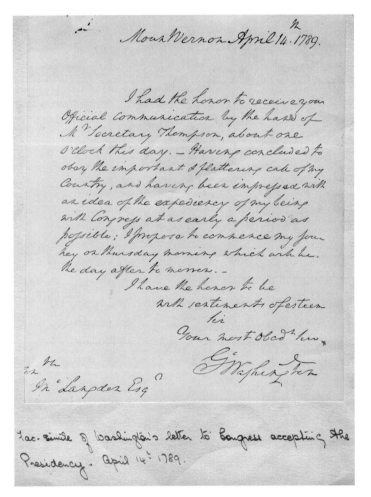

Left: A facsimile of George Washington's letter written to Congress from Mount Vernon, accepting the presidency of the United States of America.

later use to his advantage. A member of the militia, during the 1750s, Washington fought for the British in the Seven Years War (known locally as the French and Indian Wars), and acquired a reputation as a reliable and courageous commander. In 1758, he resigned his commission to return to his estates and life as a planter. His marriage to the widow Martha Dandridge in 1759 increased his lands and made him financially comfortable.

The gathering storm between the American colonists and their British overlords did not really impinge on Washington until 1769, when he called for a boycott of unpopular colonial taxes. This seemed to radicalize him and by 1775, outraged by Britain's dismissal of colonial grievances, he was prepared to fight for the rights of the colonies. He was appointed commander-in-chief in June 1775, and immediately re-

organized the American troops and acquired vital military supplies. Interestingly, he was much admired by his British opponents for his courage and concern for the welfare of his troops.

Washington's military leadership was far from an unremitting success, but he imposed discipline, had an excellent strategic overview, and his victories, notably at Trenton and Princeton, raised morale. The dismal winter at Valley Forge in 1777–78 was a low point, with a quarter of his troops left dead from disease, but the survivors emerged well trained in the spring of 1778, to find that the French had joined them as allies. The final defeat of the British at Yorktown in 1781 was Washington's victory and a credit to his decisiveness.

At the end of the war, he resigned from the army, content to return to his Mount Vernon estate. However, the early years of the American republic were politically chaotic, and Washington was regarded as a unifying influence. He reluctantly accepted the honor of becoming his country's first president in 1789 and remains the only president to receive 100% of the

Above: George Washington's home in Mount Vernon, Virginia was designated a National Historic Landmark in 1960.

Following pages: A 1939 painting by A. H. Rivey showing George Washington (1732 - 1799) arriving in New York, to take the oath to become the first President of the United States on April 23, 1789.

Electoral College's vote.

A staunch republican, Washington proved an able administrator and avoided as far as possible the lavish trappings of high office, establishing many of the customs that now define the presidency. Shrewdly, he grasped political realities without losing sight of his ideals, and never interfered with the policy-making powers that he felt the Constitution gave Congress. In foreign affairs, he insisted wisely upon a neutral course until the United States could grow stronger. After two terms in office, Washington retired in 1797. When he died a little over two years later, he was mourned around the world, in the words of Henry Lee a former comrade, "First in war, first in peace and first in the hearts of his countrymen."

Oprah Winfrey

Born 1954

American

Entertainer and philanthropist

From humble origins, Oprah Winfrey has emerged as one of the most influential people on the planet. Born to a single mother in 1954 Winfrey had an unstable and often traumatic upbringing. At the age of 13 Winfrey began to rebel against her abusive relatives and she ran away from home. After being denied entry into a juvenile detention facility because lack of

space for her, she was sent to live with her father in Nashville, Tennessee.

Under the strict parentage of her father, Vernon Winfrey, Oprah began to excel in school where she made the honors list and was eventually awarded a scholarship to Tennessee State University after winning an oratory competition. Her excellent public speaking skills landed Winfrey a job at a local radio station while she was still in high school. After entering college she accepted a job as a reporter with WTVF-TV and by the age of 19 Winfrey was the first black female anchor of the nightly news in Nashville. In 1976 Winfrey graduated from college with a degree in speech and communication and moved to Baltimore where she worked as a reporter, anchorwoman and host of a daytime talk show called "People Are Talking" for WJZ-TV for eight years.

In 1984 Winfrey was hired by Chicago station WLS-TV to help improve the ratings of that channel's struggling talk show *AM Chicago*. She was an instant smash hit and *AM Chicago* soon became the most popular talk show in Chicago. The program was expanded to an hour, renamed *The Oprah Winfrey Show* and syndicated to a national audience in 1985. In less than a year it was the most watched talk show on American Television. At the same time, just as the talk show was taking off, Winfrey was cast in Steven Spielberg's film adaptation of Alice Walker's *The Color Purple*, for which she received Golden Globe and Academy Award nominations. Winfrey would go on to establish her own production company, Harpo

Above: Oprah Winfrey addressing a rally for Democratic presidential hopeful Senator Barack Obama, December 8, 2007, in Des Moines, Iowa.

Above right: In September 2004 Oprah Winfrey was honored by the United Nations Association with its Global Humanitarian Action Award.

Opposite: Oprah poses backstage during the 30th Annual People's Choice Awards at the Pasadena Civic Auditorium, January 11, 2004.

Productions (Harpo is Oprah backwards), and she would produce not only her talk show, but films for both television and theatrical release as well. Her keen business acumen would see her rise to become America's first black female billionaire.

Oprah Winfrey has been listed as one of the most influential people in the world for the years 2004-2008 and one of the 100 most influential people of the 20th century by *Time* magazine. Her book club feature on *The Oprah Winfrey Show* regularly sees her book selection rocket to the top of the bestseller lists. She was an active supporter of Barack Obama and it is estimated she may have been responsible for pulling in one million votes for the Democratic candidate. She donated $40 million to establish *The Oprah Winfrey Leadership Academy for Girls* near Johannesburg, South Africa. Oprah is a truly remarkable media phenomenon and one of only a handful of celebrities who is recognized around the world by simply one name.

Frank Lloyd Wright

1867–1959
American
Architect

In 1991 the American Institute of Architects declared Frank Lloyd Wright "the greatest American architect of all time." With his distinctive, rather minimalist style, he changed the way homes are designed, using interior space as a flowing, flexible living area instead of individual box-like rooms. His advocacy of organic architecture that merges into the landscape remains very influential, and his iconic Prairie Style buildings are generally regarded as his signature style.

Over his long career Wright designed over 1,000 projects, about half of which were built. A number of these have collapsed or been demolished including his most important foreign commission, the Imperial Hotel in Tokyo, which was demolished in 1968.

Wright was an egoist and demanded total control over his projects: he would only accept commissions from clients who understood this and would follow his directions. He also liked to design the interior — lighting, furniture, art, glass and

Above: Frank Lloyd Wright, photographed in January 1950, the year that the Thomas E. Keys Residence that he designed was built in Rochester, Minnesota.

Left: A view of the central dome and interior walkways of the Solomon R. Guggenheim Museum in New York. This was Wright's last building.

carpets. Wright was the greatest advocate of his own work and spent considerable energy and time writing and lecturing about his theories and works.

Frank Lloyd Wright was born in Richland Center, in rural Wisconsin in 1867. His fearsome and hugely ambitious mother decided he would become an architect and put all her energies into making it happen. After briefly attending the University of Wisconsin, Wright moved to Chicago where he joined Adler and Sullivan, the leading architectural Modernist firm. Apprenticed to Louis Sullivan, he stayed for six years until he fell out with his mentor for "bootlegging" his own residential commissions.

Above: Fallingwater (also known as the Edgar J. Kaufmann Sr. Residence) in Bear Run, Pennsylvania, is one of Wright's most iconic buildings.

Wright went into private practice from his home in Oak Park Illinois and became a much sought-after architect, particularly for residential homes in his immediate locality. Between 1900 and 1917 Wright produced his Prairie Houses — such as the Robie House and Darwin D. Martin House — which were characterized by strong horizontal elevations that emphasized a long, low building, topped by a shallow sloping roof with deep overhangs and terraces; inside they were open plan and full of natural light. His larger early commercial commissions include the Unity Temple in Oak Park and the Larkin Administration Building (now demolished) in Buffalo, New York.

Wright lived a colorful personal life that alienated many Americans — he left his wife and six children when, in 1909, he ran off with Mamah Cheney, the wife of a client, which necessitated a year-long stay in Europe. Throughout his life he led an extravagant lifestyle and was constantly in debt. Wright returned to the US in 1910 and built Taliesin, his home in Wisconsin. In 1914 a deranged servant set fire to Taliesin causing the death of Mamah Cheney, her two children and four others.

Between 1935 and 1939 Wright built his most famous residence, Fallingwater near Pittsburgh. The romance of the building lay in the way it merged into the surrounding woodland with a stream cascading from underneath the cantilevered balcony. In the 1930s Wright designed the Unison houses, which were smaller and simpler economic homes designed for the middle classes, but they were still beyond the pocket of most people.

Wright's last major project was the spiral-shaped Solomon R. Guggenheim Museum in New York, which occupied a difficult site on Fifth Avenue and took him 16 years to complete.

Orville and Wilbur Wright

Orville 1871–1948

Wilbur 1867–1912

American

Pioneer aeronautic engineers

On December 17, 1903 the brothers Orville and Wilbur Wright were taking turns to test their flying machine, the *Flyer*, at Kitty Hawk, North Carolina. It was Orville's turn when the machine lurched 10 feet into the air and traveled 120 feet at a speed of 6.8 mph in 12 seconds before coming back to ground. They made several further flights that day and then sent a telegram to their father, telling him to "inform the press." After years of experimentation, they had made man's first controlled, heavier than air, powered flight. Their plane, the *Flyer I*, is now exhibited at the National Air and Space Museum, Washington DC.

In 1889 Orville opened a print shop, having designed and built his own printing press – an early indication of his remarkable technical skills. In 1892, taking advantage of the recent craze for cycling, the brothers opened the Wright Cycle Exchange (later changed to the Wright Cycle Company), a bicycle sales and repair shop, and in 1896 started manufacturing their own brand of bikes. Inspired by newspaper articles about the glider pioneer Otto Lilienthal in Germany, they started investigating the possibilities of flight and began actual mechanical experimentation in 1899, convinced that aspects of bicycle technology (which required lightweight tubular materials) could be adapted to the principles of aeronautical mechanics.

Over the next two years the Wright brothers worked to improve their flying machine and developed a fixed-wing aircraft. They analyzed the mechanical problem and realized that a successful flying machine would need wings to achieve lift and a propulsion system to move it through the air. Other pioneers were working to achieve man-powered flight, but the Wright brothers were the first to devise the controls that made flight possible. The brothers believed the key to successful flight lay with complete pilot control, and unlike other pioneers, made this the basis of their experiments. From 1900 they experimented with gliders — conducting wind tunnel tests that allowed them to design much more effective propellers

Below: Orville Wright at the controls of the very first sustained, controlled flight by a heavier-than-air craft. Wilbur Wright is running alongside.

Above: Wilbur Wright at the controls of a plane at Auvours, near Paris, c.1909. He died three years later from typhoid.

Right: Aviation pioneers Orville (left) and his brother Wilbur, c. 1910. They started by building bicycles and moved on to building heavier-than-air flying machines

and wings — to understand and learn how to control flight, before even attempting powered flight. They chose Kitty Hawk in South Carolina, a windswept village of tall dunes and sand soft enough to absorb their impact on landing.

Their breakthrough was the "three-axis control," which allowed the pilot to steer and balance the craft. Their device became the standard for fixed wing planes and in essence remains the same even today. They achieved four short flights on December 17 1903, which were witnessed by five onlookers. In 1904 and 1905 they built two further aircraft, gradually improving the design so that by 1905, they could remain airborne for up to 39 minutes.

In 1909 they formed the Wright Company, with Wilbur as president and Orville vice-president having signed a contract with the US Army the previous year. They opened an airplane factory in Dayton, Ohio and a flying school at Huffman Prairie. Wilbur traveled to France in 1908, where he made the first European flight at Le Mans, France.

With Wilbur's death from typhoid in 1912, Orville remained to continue their work in aeronautics and became one of the most celebrated Americans of his age.

Boris Yeltsin

1931–2007

Russian

Politician and President of Russia, 1990–1999

Boris Yeltsin was an often-unpredictable politician who became the first popularly elected leader in Russian history. He faced the formidable task of moving the country's Communist, state-run infrastructure to a democratic, free-market economy and by the time of his resignation eight years later, had done much to achieve this.

Yeltsin was born in Sverdlovsk, an industrial town in the east foothills of the Ural Mountains and was educated at the Ural Polytechnic. He began his working life in the construction industry but joined the Communist Party of the Soviet Union

Below: Acting President Vladimir Putin and retiring President Boris Yeltsin shake hands as Yeltsin leaves the Kremlin on December 31, 1999.

(CPSU) in 1961. He steadily rose through the local party hierarchy and was made first secretary of the Sverdlovsk region in 1976. He came to the attention of the reform-minded premier Mikhail Gorbachev, who was attempting to modernize the Soviet Union through the processes of glasnost (openness) and perestroika (restructuring), and was made a member of the CPSU's Central Committee in April 1985. He worked briefly with Nikolai Ryzhkov, who had studied at the same polytechnic and was the recently appointed secretary for the economy.

This appointment was brief and in the same year Yeltsin was made the party chief in Moscow, replacing the disgraced Viktor Grishin. Yeltsin soon acquired a reputation as a blunt, no-nonsense politician and began to root out corruption in the CPSU Moscow machine. He was elected to the party's politburo, but his outspokenness got the better of him the following year. During a meeting of the party's Central Committee, he heavily criticized certain conservative members for their attempts to sabotage the reform program and was downgraded to an insignificant post for his troubles.

Yeltsin's career revived in 1989 when he was elected to the newly created Congress of the USSR's People's Congress and in June 1991 he was elected president of the Russian republic. His greatest moment came in August 1991, when he was largely instrumental in stopping a coup against Gorbachev (by then the USSR president) by various reactionary elements within the Communist Party. Yeltsin took centre stage in the highly publicized event in Moscow, not least because Gorbachev was being held incommunicado by the plotters. Gorbachev was fatally undermined by the coup and, after 11 former Soviet republics formed the Commonwealth of Independent States in December in a move that effectively abolished his position as president of the Soviet Union, he resigned. Yeltsin was therefore his country's leading politician

Yeltsin himself survived a coup in late 1993 and the following year used force to prevent Chechnya from breaking away

Above: President Yeltsin calling on supporters for a general strike and to resist the pro-Communist coup against Soviet President Gorbachev.

Right: Boris Yeltsin taking the oath of the presidency at the Supreme Soviet in Moscow in July 1991. He was the first democratically elected Russian leader.

from the federation in what would become a costly and unpopular war. Various other domestic problems, not least a severe economic downturn and dangerously high levels of both petty and organized crime also beset the president. By 1998 the economy was in freefall and Yeltsin responded by sacking his government on two occasions. He was also gaffe-prone and suffered from ill health, in part because of his alcohol problems. With his popularity rating at rock bottom, Yeltsin resigned on 31 December 1999 and was replaced by Vladimir Putin. Nevertheless, he had played a large part in the dissolution of the Soviet Union and the end of Communist rule, thereby allowing democracy to take hold in his country.

PICTURE CREDITS

The Aerospace Corporation
Page 76 (both),

**Australian Government Department of Defence - Defence Science
and Technology Organisation**
Page 208,

Corbis Images
218, (below) Duncan Andison/Corbis,

Getty Images
Front end papers Alex Wong/Getty Images; pages 1 AFP/Getty
Images; 2 NASA/Central Press/Getty Images; Pages 7 Time Life
Pictures/Mansell/Time Life Pictures/Getty Image; 8 Hulton
Archive/Getty Images; 9 (top right) AFP/Getty Images; 9 (left &
below right); 10; 11 (both) AFP/Getty Images; 12 (below) The
Bridgeman Art Library/Getty Images; 12 (top), 13 (both) Hulton
Archive/Getty Images; 14; 15 (top) Hulton Archive/Getty Images;
16, 17 (both) Hulton Archive/Getty Images; 18 (top) Michael
Ochs Archives/Getty Images; 18 (below); 19 (top) Redfern/Getty
Images; 19 (below), 20, 21 (both) Hulton Archive/Getty Images;
22; 23 (both) AFP/Getty Images; 24 (both), 25 Hulton Archive/
Getty Images; 26 (top) The Bridgeman Art Library/Getty Images;
26 (below), 27, Hulton Archive/Getty Images; 28 (both) AFP/
Getty Images; 29 (both), 30, 31 (below) Hulton Archive/Getty
Images; 31 (top) AFP/Getty Images; 32 The Bridgeman Art Library/
Getty Images; 33 (both), 34 (both), 35 (both) Hulton Archive/
Getty Images; 38 AFP/Getty Images; 39 (top), 40, 41 (all 3), 42
(top) Hulton Archive/Getty Images; 42 (below left & right), 43
AFP/Getty Images; 44, 45 (both), 46 (both), 47 (below right)
Hulton Archive/Getty Images; 47 (top) AFP/Getty Images; 47
(below right); 48, 49 (both) AFP/Getty Images; 50, 51 (below
left) Hulton Archive/Getty Images; 51 (below right) AFP/Getty
Images; 51 (top); 52 Michael Ochs Archives/Getty Images; 53 (all
3); 54, 55 (both), 56, 57 (top) Hulton Archive/Getty Images; 57
(below) Popperfoto/Getty Images; 58 (both), 59, 60 (both), Hulton
Archive/Getty Images; 61 Popperfoto/Getty Images; 62 (both),
63, 64, 65 (both), 66 (both) Hulton Archive/Getty Images; 67
The Bridgeman Art Library/Getty Images; 68 AFP/Getty Images;
69 (both) Hulton Archive/Getty Images; 70 Time & Life Pictures/
Getty Images; 71 Hulton Archive/Getty Images; 72, 73 (both) AFP/
Getty Images; 74, 75 (below) Hulton Archive/Getty Images; 75
AFP/Getty Images; 77, 78 (top) AFP/Getty Images; 78 (below); 79
AFP/Getty Images; 80, 81 (both), 82 (both), 83 Hulton Archive/
Getty Images; 84 (top); 84 (below), 85 (top) Hulton Archive/
Getty Images; 85 (below) AFP/Getty Images; 86 (below) Hulton
Archive/Getty Images; 86 (top) AFP/Getty Images; 87, 88, 89
(both) Hulton Archive/Getty Images; 90 (top) Time & Life
Pictures/Getty Images; 90 (below) Roger Viollet/Getty Images; 91;
92, 93 (top) AFP/Getty Images; 93 (below left & right); 94 , 95
(below) AFP/Getty Images; 95 (top); 96 (both), 97 (both) AFP/
Getty Images; 98 (both), 99 (below) AFP/Getty Images; 99 (top)
Time & Life Pictures/Getty Images; 100, 101 (top & below right)
AFP/Getty Images; 101 (below left); 102; 103 (top) AFP/Getty
Images; 104, 105 (both) AFP/Getty Images; 106 (both), 107 (both),
108, 109 (both), Hulton Archive/Getty Images; 110, 111 (below)
Hulton Archive/Getty Images; 111 (top) Time & Life Pictures/
Getty Images; 112, 113 (below) Hulton Archive/Getty Images;
114 (both); 115 AFP/Getty Images; 116 Time & Life Pictures/
Getty Images; 117 (top) AFP/Getty Images; 117 (below), 118 (top)
Hulton Archive/Getty Images; 118 (below); 119 (top) AFP/Getty
Images; 119 below; 120, 121 (both), 122 (both) Hulton Archive/
Getty Images; 123; 124, 125 Time & Life Pictures/Getty Images;
126, 127 (top) Hulton Archive/Getty Images; 127 (below) Time
& Life Pictures/Getty Images; 128 Hulton Archive/Getty Images;
129 (both) Time & Life Pictures/Getty Images; 130, 131 (both),
132 (below) Hulton Archive/Getty Images; 132 (top), 133 (both)
Archive/Getty Images; 134, 135 (top right) Michael Ochs Archives/
Getty Images; 135 (below) Hulton Archive/Getty Images; 135
(top left), 136 (below) AFP/Getty Images; 136 (top); 137 (both)
AFP/Getty Images; 138, 140, 141 (top) Hulton Archive/Getty
Images; 141 (below) AFP/Getty Images; 142, 143 (below) Hulton
Archive/Getty Images; 143 (top); 144, 145 (both), 146 (both), 147
Hulton Archive/Getty Images; 148 AFP/Getty Images; 149 (top);
149 (below) Hulton Archive/Getty Images; 150, 151 (both) 152
(both), 153 (both) Hulton Archive/Getty Images; 154 (both) AFP/

INDEX